Universitext

Springer
Berlin
Heidelberg
New York
Hong Kong
London
Milan
Paris
Tokyo

Vladimir I. Arnold

Lectures
on Partial Differential
Equations

Translated by Roger Cooke

 Springer

PHASIS

Vladimir I. Arnold

Steklov Mathematical Institute
ul. Gubkina 8
117966 Moscow, Russia
e-mail: arnold@genesis.mi.ras.ru
and
CEREMADE
Université de Paris-Dauphine
Place du Maréchal de Lattre de Tassigny
75775 Paris Cedex 16, France
e-mail: arnold@ceremade.dauphine.fr

Roger Cooke (Translator)

Department of Mathematics and Statistics
University of Vermont
16 Colchester Ave.
Burlington, VT 05401, USA
e-mail: cooke@emba.uvm.edu

Cataloging-in-Publication Data applied for
A catalog record for this book is available from the Library of Congress.
Bibliographic information published by Die Deutsche Bibliothek
Die Deutsche Bibliothek lists this publication in the Deutsche Nationalbibliografie;
detailed bibliographic data is available in the Internet at http://dnb.ddb.de

Originally published in Russian as "Lektsii ob uravneniyakh s chastnymi proizvodnymi" by PHASIS, Moscow, Russia, 1997 (ISBN 5-7036-0035-9)

Mathematics Subject Classification (2000): 35-01, 70-01

ISBN 3-540-40448-1 Springer-Verlag Berlin Heidelberg New York

Springer-Verlag Berlin Heidelberg New York
A part of Springer Science+Business Media GmbH

http://www.springer.de

© Springer-Verlag Berlin Heidelberg and PHASIS Moscow 2004
Printed in Germany

The use of general descriptive names, registered names, trademarks, etc. in this publication does not imply, even in the absence of a specific statement, that such names are exempt from the relevant protective laws and regulations and therefore free for general use.

Cover design: *design & production* GmbH, Heidelberg
Edited by PHASIS using a Springer LaTeX macro package

Printed on acid-free paper   46/3111 - 5 4 3 2 1   Spin 1 1 0 1 9 3 2 9

Preface to the Second Russian Edition

In the mid-twentieth century the theory of partial differential equations was considered the summit of mathematics, both because of the difficulty and significance of the problems it solved and because it came into existence later than most areas of mathematics.

Nowadays many are inclined to look disparagingly at this remarkable area of mathematics as an old-fashioned art of juggling inequalities or as a testing ground for applications of functional analysis. Courses in this subject have even disappeared from the obligatory program of many universities (for example, in Paris). Moreover, such remarkable textbooks as the classical three-volume work of Goursat have been removed as superfluous from the library of the University of Paris-7 (and only through my own intervention was it possible to save them, along with the lectures of Klein, Picard, Hermite, Darboux, Jordan, ...).

The cause of this degeneration of an important general mathematical theory into an endless stream of papers bearing titles like "On a property of a solution of a boundary-value problem for an equation" is most likely the attempt to create a unified, all-encompassing, superabstract "theory of everything."

The principal source of partial differential equations is found in the continuous-medium models of mathematical and theoretical physics. Attempts to extend the remarkable achievements of mathematical physics to systems that match its models only formally lead to complicated theories that are difficult to visualize as a whole, just as attempts to extend the geometry of second-order surfaces and the algebra of quadratic forms to objects of higher degrees quickly leads to the detritus of algebraic geometry with its discouraging hierarchy of complicated degeneracies and answers that can be computed only theoretically.

The situation is even worse in the theory of partial differential equations: here the difficulties of commutative algebraic geometry are inextricably bound up with noncommutative differential algebra, in addition to which the topological and analytic problems that arise are profoundly nontrivial.

At the same time, general physical principles and also general concepts such as energy, the variational principle, Huygens' principle, the Lagrangian, the Legendre transformation, the Hamiltonian, eigenvalues and eigenfunctions, wave-particle duality, dispersion relations, and fundamental solutions interact elegantly in numerous highly important problems of mathematical physics. The study of these problems motivated the development of large areas of mathematics such as the theory of Fourier series and integrals, functional analysis, algebraic geometry, symplectic and contact topology, the theory of asymptotics of integrals, microlocal analysis, the index theory of (pseudo-)differential operators, and so forth.

Familiarity with these fundamental mathematical ideas is, in my view, absolutely essential for every working mathematician. The exclusion of them from the university mathematical curriculum, which has occurred and continues to occur in many Western universities under the influence of the axiomaticist/scholastics (who know nothing about applications and have no desire to know anything except the "abstract nonsense" of the algebraists) seems to me to be an extremely dangerous consequence of Bourbakization of both mathematics and its teaching. The effort to destroy this unnecessary scholastic pseudoscience is a natural and proper reaction of society (including scientific society) to the irresponsible and self-destructive aggressiveness of the "superpure" mathematicians educated in the spirit of Hardy and Bourbaki.

The author of this very short course of lectures has attempted to make students of mathematics with minimal knowledge (linear algebra and the basics of analysis, including ordinary differential equations) acquainted with a kaleidoscope of fundamental ideas of mathematics and physics. Instead of the principle of maximal generality that is usual in mathematical books the author has attempted to adhere to the principle of minimal generality, according to which every idea should first be clearly understood in the simplest situation; only then can the method developed be extended to more complicated cases.

Although it is usually simpler to prove a general fact than to prove numerous special cases of it, for a student the content of a mathematical theory is never larger than the set of examples that are thoroughly understood. That is why it is examples and ideas, rather than general theorems and axioms, that form the basis of this book. The examination problems at the end of the course form an essential part of it.

Particular attention has been devoted to the interaction of the subject with other areas of mathematics: the geometry of manifolds, symplectic and contact geometry, complex analysis, calculus of variations, and topology. The author has aimed at a student who is eager to learn, but hopes that through this book even professional mathematicians in other specialties can become acquainted with the basic and therefore simple ideas of mathematical physics and the theory of partial differential equations.

The present course of lectures was delivered to third-year students in the Mathematical College of the Independent University of Moscow during the fall semester of the 1994/1995 academic year, Lectures 4 and 5 having been delivered by Yu. S. Il'yashenko and Lecture 8 by A. G. Khovanskiĭ. All the lectures were written up by V. M. Imaĭkin, and the assembled lectures were then revised by the author. The author is deeply grateful to all of them.

The first edition of this course appeared in 1995, published by the press of the Mathematical College of the Independent University of Moscow. A number of additions and corrections have been made in the present edition.

Contents

Lecture 1

The General Theory for One First-Order Equation

In contrast to ordinary differential equations, there is no unified theory of partial differential equations. Some equations have their own theories, while others have no theory at all. The reason for this complexity is a more complicated geometry. In the case of an ordinary differential equation a locally integrable vector field (that is, one having integral curves) is defined on a manifold. For a partial differential equation a subspace of the tangent space of dimension greater than 1 is defined at each point of the manifold. As is known, even a field of two-dimensional planes in three-dimensional space is in general not integrable.

Example. In a space with coordinates x, y, and z we consider the field of planes given by the equation $\mathrm{d}z = y\,\mathrm{d}x$. (This gives a linear equation for the coordinates of the tangent vector at each point, and that equation determines a plane.)

Problem 1. Draw this field of planes and prove that it has no integral surface, that is, no surface whose tangent plane at every point coincides with the plane of the field.

Thus integrable fields of planes are an exceptional phenomenon.

An *integral submanifold* of a field of tangent subspaces on a manifold is a submanifold whose tangent plane at each point is contained in the subspace of the field. If an integral submanifold can be drawn, its dimension usually does not coincide with that of the planes of the field.

In this lecture we shall consider a case in which there is a complete theory, namely the case of one first-order equation. From the physical point of view this case is the duality that occurs in describing a phenomenon using waves or particles. The field satisfies a certain first-order partial differential equation, the evolution of the particles is described by ordinary differential equations, and there is a method of reducing the partial differential equation to a system of ordinary differential equations; in that way one can reduce the study of wave propagation to the study of the evolution of particles.

We shall write everything in a local coordinate system: $x = (x_1, \ldots, x_n)$ are the coordinates (independent variables), $y = u(x)$ is an unknown function of the coordinates. The letter y by itself denotes the coordinate on the axis of values, and we denote the partial derivatives by the letter p: $p_i = \frac{\partial u}{\partial x_i} = u_{x_i}$.

The general first-order partial differential equation has the form

$$F(x_1, \ldots, x_n, y, p_1, \ldots, p_n) = 0 .$$

Examples.

$$\frac{\partial u}{\partial x_1} = 0 ; \tag{1.1}$$

$$\left(\frac{\partial u}{\partial x_1} \right)^2 + \left(\frac{\partial u}{\partial x_2} \right)^2 = 1 \tag{1.2}$$

(the eikonal equation in geometric optics);

$$u_t + u u_x = 0 \tag{1.3}$$

(Euler's equation).

Consider a convex closed curve in the plane with coordinates x_1, x_2. Outside the region bounded by the curve we consider the function u whose value at each point is the distance from that point to the curve. The function u is smooth.

Theorem 1. *The function u satisfies the equation* (1.2).

PROOF. Equation (1.2) says that the square-norm of the gradient of u equals 1. We recall the geometric meaning of the gradient. It is a vector pointing in the direction of maximal rate of increase of the function, and its length is that maximal rate of increase. The assertion of the theorem is now obvious. □

Problem 2. a) Prove that any solution of the equation (1.2) is locally the sum of a constant and the distance to some curve.

b) Understand where the wave-particle duality occurs in this situation. (In case of difficulty see below p. 14, Fig. 2.2.)

Consider the field $u(t, x)$ of velocities of particles moving freely along a line (Fig. 1.1). The law of free motion of a particle has the form $x = \varphi(t) = x_0 + vt$, where v is the velocity of the particle. The function φ satisfies Newton's equation $\frac{d^2 \varphi}{dt^2} = 0$. We now give a description of the motion in terms of the velocity field u: by definition $\frac{d\varphi}{dt} = u(t, \varphi(t))$. We differentiate with respect to t, obtaining the Euler equation:

$$\frac{d^2 \varphi}{dt^2} = u_t + u_x u = 0 .$$

Fig. 1.1. A particle on a line

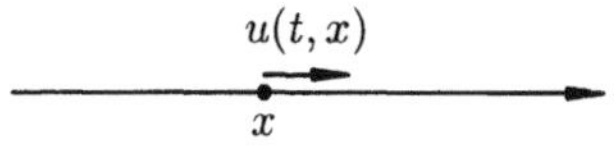

$$u(t,x)$$

$$\xrightarrow{\qquad\qquad\bullet\quad\rightarrow\qquad\qquad\qquad}$$
$$x$$

Conversely, Newton's equation can be derived from Euler's, that is, these descriptions of the motion using Euler's equation for a field and Newton's equation for particles are equivalent. We shall also construct a procedure for the general case that makes it possible to reduce equations for waves to equations for the evolution of particles. First, however, we consider some simpler examples of linear equations.

1. Let $v = v(x)$ be a vector field on a manifold or in a region of Euclidean space. Consider the equation $L_v(u) = 0$, where the operator L_v denotes the derivative in the direction of the vector field (the Lie derivative).

In coordinates this equation has the form $v_1 \frac{\partial u}{\partial x_1} + \cdots + v_n \frac{\partial u}{\partial x_n} = 0$; it is called a *homogeneous linear first-order partial differential equation.*

For the function u to be a solution of this equation it is necessary and sufficient that u be constant along the phase curves of the field v. Thus *the solutions of our equation are the first integrals of the field.*

For example, consider the field $v = \sum_{i=1}^{n} x_i \frac{\partial}{\partial x_i}$ in Fig. 1.2. Let us solve the equation $L_v(u) = 0$ for this field v. The phase curves are the rays $x = e^t x_0$ emanating from the origin. The solution must be constant along each such ray. If we require continuity at the origin, we find that the only solutions are constants. The constants form a one-dimensional vector space. (The solutions of a linear equation necessarily form a vector space.)

Fig. 1.2. An Eulerian field

In contrast to this example, the solutions of a linear partial differential equation in general form an infinite-dimensional space. For example, for the equation $\frac{\partial u}{\partial x_1} = 0$ the solution space coincides with the space of functions of $n - 1$ variables:

$$u = \varphi(x_2, \ldots, x_n) \, .$$

It turns out that the same is true for an equation in general position in a neighborhood of a regular point.

The Cauchy Problem. Consider a smooth hypersurface Γ^{n-1} in x-space. The *Cauchy problem* is the following: *find a solution of the equation $L_v(u) = 0$ that coincides with a given function on this hypersurface* (Fig. 1.3).

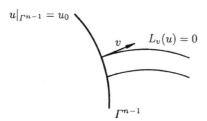

$u|_{\Gamma^{n-1}} = u_0$

$v \qquad L_v(u) = 0$

Γ^{n-1}

Fig. 1.3. The Cauchy problem

A point of the hypersurface is called *noncharacteristic* if the field v is transversal to the surface at that point.

Theorem 2. *The Cauchy problem has a unique solution in a neighborhood of each noncharacteristic point.*

PROOF. Using a smooth change of variables we can rectify the vector field and convert Γ into the hyperplane $x_1 = 0$. Then in a small neighborhood of a noncharacteristic point we obtain the following problem:

$$\frac{\partial u}{\partial x_1} = 0 \,, \qquad u\big|_{0, x_2, \ldots, x_n} = u_0(x_2, \ldots, x_n) \,,$$

which has a unique solution.   □

2. Consider the Cauchy problem for a more general, *inhomogeneous linear equation*:
$$L_v(u) = f \,, \qquad u\big|_{\Gamma^{n-1}} = u_0 \,.$$

The solutions of such a problem form an *affine* space. (The general solution of the inhomogeneous equation is the sum of the general solution of the homogeneous equation and a particular solution of the inhomogeneous equation.)

By a smooth change of variable the problem can be brought into the form

$$\frac{\partial u}{\partial x_1} = f(x_1, x_2, \ldots, x_n) \,, \qquad u\big|_{0, x_2, \ldots, x_n} = u_0(x_2, \ldots, x_n) \,.$$

This problem has a unique solution:

$$u(x_1, \ldots) = u_0(\ldots) + \int_0^{x_1} f(\xi, \ldots) \, \mathrm{d}\xi \,.$$

3. An equation that is linear with respect to the derivatives is *quasilinear.* In coordinates a first-order quasilinear equation has the form

$$a_1(x, u)\frac{\partial u}{\partial x_1} + \cdots + a_n(x, u)\frac{\partial u}{\partial x_n} = f(x, u) . \tag{*}$$

We remark that in the first two cases the field v is invariantly (independently of the coordinates) connected with the differential operator. How can a geometric object be invariantly connected with a quasilinear equation?

Consider the space with coordinates $(x_1, \ldots, x_n, y)$, the space of *0-jets of functions* of $(x_1, \ldots, x_n)$, which we denote $J^0(\mathbb{R}^n, \mathbb{R})$ or, more briefly, J^0.

We recall that the *space of k-jets* of functions of $(x_1, \ldots, x_n)$ is the space of Taylor polynomials of degree k.

We note that the argument of $(x_1, \ldots, x_n, y, p_1, \ldots, p_n)$ in a first-order equation is the 1-jet of the function. Thus a first-order equation can be interpreted as a hypersurface in the space $J^1(\mathbb{R}^n, \mathbb{R})$ of 1-jets of functions. The space of 1-jets of real-valued functions of n variables can be identified with a $(2n + 1)$-dimensional space: $J^1(\mathbb{R}^n, \mathbb{R}) \approx \mathbb{R}^{2n+1}$. For example, for functions on the plane we obtain a five-dimensional space of 1-jets.

The solution of the equation $(*)$ can be constructed using its characteristics (curves of a special form in J^0). The word "characteristic" in mathematics always means "invariantly connected." For example, the characteristic polynomial of a matrix is invariantly connected with an operator and independent of the basis in which the matrix is formed. The characteristic subgroups of a group are those that are invariant with respect to the automorphisms of the group. The characteristic classes in topology are invariant with respect to suitable mappings.

The vector field v (in the space of independent variables) is called the *characteristic field* of the linear equation $L_v(u) = f$.

Definition. The *characteristic field* of the quasilinear equation $(*)$ is the vector field A in J^0 with components $(a_1, \ldots, a_n, f)$.

Claim. *The direction of this field is characteristic.*

Indeed, let u be a solution. Its graph is a certain hypersurface in J^0. This hypersurface is tangent to the field A, as the equation itself asserts. The converse is also true: if the graph of a function is tangent to the field A at each point, then the function is a solution.

The method of solving a quasi-linear equation becomes clear from this. We draw the phase curves of the characteristic field in J^0. They are called characteristics. If a characteristic has a point in common with the graph of a solution, it lies entirely on that graph. Thus the graph is composed of characteristics.

The *Cauchy problem for a quasilinear equation* is stated in analogy with the preceding cases. To be specific, suppose given a smooth hypersurface Γ^{n-1}

in x-space and an initial function u_0 defined on the hypersurface. The graph of this function is a surface $\widehat{\Gamma}$ in J^0, which we regard as the initial submanifold of Fig. 1.4.

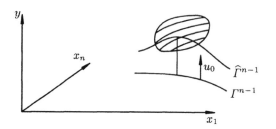

Fig. 1.4. Characteristics of a quasilinear equation passing through the initial manifold $\widehat{\Gamma}^{n-1}$

If the characteristics are not tangent to the hypersurface $\widehat{\Gamma}$, the graph of the solution is composed locally of them.

In this case two conditions are needed for a point to be noncharacteristic: the field A must not be tangent to $\widehat{\Gamma}^{n-1}$ and, in order for an actual graph to result, the vector of the field must not be vertical, that is, the component a must not be zero.

Points where $a = 0$ are singular; at these points the differential equation vanishes, becoming an algebraic equation.

Example. For the Euler equation $u_t + uu_x = 0$ the equation of the characteristics is equivalent to Newton's equation: $\dot{t} = 1$, $\dot{x} = u$, $\dot{u} = 0$.

Let us now pass to the general first-order equation.

Consider the space of 1-jets $J^1(\mathbb{R}^n, \mathbb{R})$. Instead of $\mathbb{R}^n$ one can consider an n-dimensional manifold B^n; in that case we obtain the space $J^1(B^n, \mathbb{R})$. Let (x, y, p) be local coordinates in this space.

A *first-order partial differential equation* is a smooth surface in J^1: $\Gamma^{2n} \subset J^1$.

For example, when $n = 1$ we obtain an implicit ordinary differential equation (not solved with respect to the derivative).

It turns out that there is a remarkable geometric structure in our space J^1, an invariantly defined distribution of $2n$-dimensional hypersurfaces. For example, when $n = 1$ we obtain a field of planes in three-dimensional space. The structure arises only because the space is a space of 1-jets. An analogous structure appears in spaces of jets of higher order, where it is called a *Cartan distribution*.

Each function in the space of k-jets has a k-*graph*. For a 0-jet this is the usual graph – the set of 0-jets of the function: $\Gamma_u = \left\{ j_x^0 u : x \in \mathbb{R}^n \right\} = \left\{ (x, y) : y = u(x) \right\}$. In the case of 1-jets a point of the 1-*graph* consists of the argument, the value of the function, and the values of the first-order partial derivatives: $\left\{ j_x^1 u : x \in \mathbb{R}^n \right\} = \left\{ (x, y, p) : y = u(x), p = \frac{\partial u}{\partial x} \right\}$ (see Fig. 1.5

for $n = 1$). We remark that the 1-graph is a section of the bundle over the domain of definition.

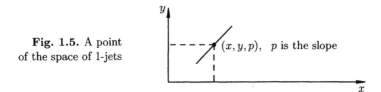

Fig. 1.5. A point of the space of 1-jets

(x, y, p), p is the slope

Remark. The surface of the 1-graph is diffeomorphic to the domain of definition of the function, x is the n-dimensional coordinate on this surface. The smoothness of this surface is 1 less than the smoothness of the function, but smoothness is preserved for an infinitely differentiable or analytic function.

Consider the tangent plane to the 1-graph. This is an n-dimensional plane in a $(2n + 1)$-dimensional space.

Theorem 3. *All tangent planes of all 1-graphs at a given point lie in the same hyperplane.*

PROOF. Along any tangent plane we have $dy = \sum \frac{\partial u}{\partial x_i} dx_i = \sum p_i \, dx_i$, or $dy = p \, dx$. Since p is fixed at a given point of the space of 1-jets, we obtain an equation for the components of the tangent vector that determines the hyperplane. Thus the tangent plane to any 1-graph lies in this hyperplane. □

For example, when $n = 1$ the equation $dy = p \, dx$ defines a vertical plane in space with the coordinates x, y, p. The tangents to the 1-graphs are the non-vertical lines lying in this plane (Fig. 1.6).

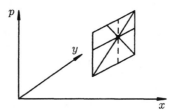

Fig. 1.6. The contact plane in the space of 1-jets

In this case one can see that the hypersurface itself is the closure of the union of the tangents to all 1-graphs passing through the given point.

Problem 3. Prove that this is true for any dimension.

Corollary. *The field of hyperplanes* $dy = p\,dx$ *constructed is invariantly determined, that is, in other coordinates it is given by the equation* $d\tilde{y} = \tilde{p}\,d\tilde{x}$.

Definition. The field of hyperplanes in J^1 is a *Cartan distribution* or a *standard contact structure*.

Problem 4. What are the dimensions of the integral manifolds for the field of contact planes? (A manifold is called integral if the tangent plane at each of its points is a subspace of the contact plane.)

A 1-graph is always an integral manifold, that is, a 1-dimensional integral manifold exists. But are there integral manifolds of higher dimension?

ANSWER. A field of contact planes has no integral manifolds of dimension greater than half the dimension of the contact plane.

Definition. An integral submanifold of a field of contact planes whose dimension is maximal (that is, equal to half the dimension of the contact plane) is a *Legendre manifold*.

For example, 1-graphs are Legendre manifolds.

Let us now return to the differential equation.

An equation is a $2n$-dimensional submanifold Γ^{2n} in J^1.

At each regular point of this surface one can distinguish a characteristic direction determined by the surface and the contact structure. We shall construct the characteristics (integral curves of this direction field), and then use them to form the integral manifolds.

At a point of the surface Γ^{2n} we consider the intersection of the tangent and contact planes. These planes either coincide or have a $(2n-1)$-dimensional intersection. In the first case the point is *singular*, in the second case it is *regular*.

We remark that for a surface Γ in general position the singular points are isolated. Indeed there are $2n$ coordinates on Γ. Consider the normal to the tangent plane and the normal to the contact plane. A point is singular if these normals have the same direction. This means that $2n$ functions of $2n$ variables vanish simultaneously. In general position this can happen only at isolated points.

Thus, at regular points there are $(2n-1)$-dimensional intersections of the tangent and contact planes. These are lines when $n = 1$, but not for $n > 1$. How can we distinguish a one-dimensional direction?

In local coordinates the contact field is given by the zeros of the 1-form $\alpha = dy - p\,dx$, and this form can be multiplied by any nonzero function without changing the field (contact structure).

The 2-form $\omega^2 = d\alpha$, which is the exterior differential of the form α, is no longer defined invariantly by the contact structure. However, the following proposition holds.

Proposition 1. *The form* $\omega\big|_{\alpha=0}$ *is defined invariantly up to multiplication by a nonzero number at each point.*

PROOF. Let $\tilde{\alpha} = f\alpha$. Then $d\tilde{\alpha} = df \wedge \alpha + f\,d\alpha$.

$$d\tilde{\alpha}\big|_{\alpha=0} = f\,d\alpha\big|_{\alpha=0},$$

that is, $\tilde{\omega}^2$ differs from ω^2 by multiplication by a number at each point (the *conformal type* of the form ω^2 is said to be invariantly defined). We remark that $\tilde{\alpha} = 0$ wherever $\alpha = 0$. $\square$

Proposition 2. $\omega\big|_{\alpha=0}$ *is a symplectic structure.*

We recall that a *symplectic structure* is a nondegenerate skew-symmetric bilinear form in an even-dimensional space.

Nondegeneracy of the form ω means that $\forall \xi \neq 0 \ \exists \eta \ \omega(\xi, \eta) \neq 0$.

PROOF. In local coordinates our form is written as: $d\alpha = -\sum dp_i \wedge dx_i$, where p_i and x_i are coordinates in the plane $\alpha = 0$. $\square$

Exercise. Write out the matrix of the form $\sum dp_i \wedge dx_i$ and verify that this form is nondegenerate.

This form is called the *skew-scalar product*. Let us clarify its geometric meaning.

Let $n = 1$. Then $\omega = dx \wedge dp$. The value of this form on a pair of vectors is the oriented area of the parallelogram spanned by the vectors, as in Fig. 1.7. In the case of higher dimension $\omega(\xi, \eta)$ is the sum of the oriented areas of the projections of the parallelogram with sides ξ and η on the plane with coordinates (x_i, p_i).

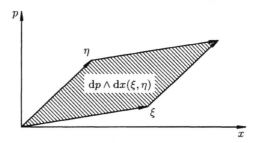

Fig. 1.7. A symplectic structure

We recall that in Euclidean space there is a concept of orthogonal complement. In n-dimensional space the orthogonal complement to a k-dimensional subspace is an $(n-k)$-dimensional subspace. The proof of this fact uses only the bilinearity and nondegeneracy of the scalar product, not its symmetry, so that the same is true in the case of the skew-scalar product.

Thus the skew-orthogonal complement to a $(2n-1)$-dimensional plane in $2n$-dimensional space is a line. But, in contrast to the Euclidean case, the line lies in this plane!

Lemma. *The skew-orthogonal complement to a hyperplane in symplectic space is a line lying in the hyperplane.*

PROOF. Let the line be spanned by the vector ξ. Its skew-orthogonal complement is the hyperplane $\{\eta : \omega(\xi, \eta) = 0\}$. The vector ξ lies in this hyperplane since $\omega(\xi, \xi) = -\omega(\xi, \xi) = 0$. $\square$

Definition. A *characteristic direction* in a contact plane is the skew-orthogonal complement to the intersection of the contact plane and the plane tangent to Γ at a regular point.

This skew-orthogonal complement is a line. Thus there is an invariant contact structure and an invariant relation of skew-orthogonality in each contact plane. Consequently a characteristic direction is invariantly distinguished at each regular point (see Fig. 1.8).

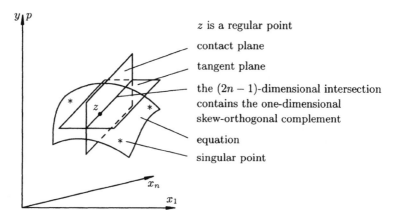

the $(2n-1)$-dimensional intersection contains the one-dimensional skew-orthogonal complement

Fig. 1.8. A characteristic direction for a general first-order equation

The integral curves of this direction field are called *characteristics*.

Problem 5. Compute the characteristic direction field in coordinates x, y, p, that is, determine it by a system of differential equations $\dot{x} = ?$, $\dot{y} = ?$, $\dot{p} = ?$.

Literature

1. Arnold, V.I.: Ordinary Differential Equations. Springer, Berlin, Chap. 2, § 11 (1992)
2. Arnold, V.I.: Geometrical Methods in the Theory of Ordinary Differential Equations, 2nd edition. Springer, New York, Chap. 2 (1988)

Lecture 2

The General Theory for One First-Order Equation (Continued)

We are considering a general first-order partial differential equation $F(x, y, p) = 0$, where $x = (x_1, \ldots, x_n)$, $p = (p_1, \ldots, p_n)$, $p_i = \frac{\partial u}{\partial x_i}$, and $y = u(x)$ is an unknown function. The equation determines a $2n$-dimensional hypersurface V^{2n} in the space J^1 of 1-jets of functions of $(x_1, \ldots, x_n)$. Each function has a 1-graph in J^1; it is a solution of the equation if its 1-graph is a submanifold in V^{2n}.

At each point of J^1 there is a contact plane K^{2n} defined in local coordinates by the equation $dy = p\, dx$; geometrically it is the closure of the union of the tangent planes to all the 1-graphs passing through the point.

At points of the surface V^{2n} the tangent plane to V^{2n} intersects the contact plane. If the intersection is $(2n-1)$-dimensional, the point is regular; otherwise it is singular. In general position singular points are isolated. At regular points z we obtain a distribution of $(2n-1)$-dimensional planes $(T_z V^{2n}) \cap K_z^{2n}$, which are subspaces of the contact planes K_z^{2n}.

Each contact plane is a symplectic space, the symplectic structure being defined by the differential 2-form $\omega^2 = d\alpha|_{K^{2n}}$, where $\alpha = dy - p\, dx$.

In coordinates we have $\omega^2 = dx \wedge dp := \sum_{i=1}^{n} dx_i \wedge dp_i$; (x, p) constitute coordinates in the contact plane.

Lemma. *The form ω^2 is nondegenerate.*

PROOF. Its matrix in the coordinates $x_1, p_1, x_2, p_2, \ldots$ has the form

$$
\begin{pmatrix}
0 & 1 & & & & 0 \\
-1 & 0 & & & & \\
& & 0 & 1 & & \\
& & -1 & 0 & & \\
0 & & & & & \ddots
\end{pmatrix}.
$$ $\qquad\square$

Hence the form ω^2 defines a skew-scalar product in K^{2n}, and the skew-orthogonal complement to each subspace in K^{2n} is defined and has the complementary dimension. In particular, the skew-orthogonal complement to the

$(2n-1)$-dimensional intersection $(T_z V^{2n}) \cap K_z^{2n}$ is one-dimensional. This direction is *characteristic for the equation*.

Problem 1. Prove that the characteristic line lies in $(T_z V^{2n}) \cap K_z^{2n}$.

Let us compute the components of the characteristic vector explicitly in terms of the equation. Let (x, y, p) be the coordinates in $T_z J^1$. The characteristic vector must be tangent to V^{2n}: by differentiating the equation, we obtain the first condition

$$F_x \dot{x} + F_y \dot{y} + F_p \dot{p} = 0 . \tag{2.1}$$

Moreover, the vector must lie in the contact plane; hence we obtain the second condition

$$\dot{y} = p\dot{x} . \tag{2.2}$$

Eliminating $\dot{y}$ from these two equations, we obtain the equation of the $(2n-1)$-dimensional intersection K^{2n-1} of the tangent plane and the contact plane:

$$(F_x + F_y p)\dot{x} + F_p \dot{p} = 0 . \tag{2.3}$$

This equation is written in (x, p)-coordinates, which we can regard as coordinates in the contact plane, since they project one-to-one onto the (x, p)-hyperplane in J^1 (see Fig. 2.1).

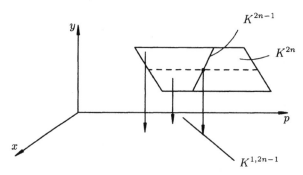

Fig. 2.1. x and p are coordinates in the contact plane

It remains for us to find the skew-orthogonal complement to K^{2n-1}. Theoretically this problem could be solved algorithmically by writing a suitable system of linear equations, but we shall take advantage of a useful observation.

As an example let us consider the case $n = 1$. Then the value of the skew-scalar product $dx \wedge dp$ on a pair of tangent vectors $(\dot{x}, \dot{p})$, (x', p') is simply the determinant of their coordinates:

$$dx \wedge dp\big((\dot{x}, \dot{p}), (x', p')\big) = \begin{vmatrix} \dot{x} & \dot{p} \\ x' & p' \end{vmatrix} = \dot{x} p' - x' \dot{p} .$$

It turns out that the same is true in general if x and p are interpreted as vectors and their product as the scalar product.

Problem 2. Prove that

$$\omega^2\big((\dot{x},\dot{p}),(x',p')\big) = \dot{x}p' - x'\dot{p}\,. \tag{2.4}$$

But let us examine (2.3) carefully. The equation itself has the form (2.4), that is, it expresses the fact that the skew-scalar product equals 0. Thus we can write out the skew-scalar complement to K^{2n-1} directly from this equation: $p' = F_x + pF_y$, $x' = -F_p$. (To be sure, it is the oppositely directed vector that is traditionally called the characteristic vector.) As a result, we have proved the following theorem.

Theorem 1. *The direction field of the vector field given by* $x' = F_p$, $p' = -(F_x + pF_y)$, $y' = pF_p$ *is characteristic (the last component being obtained from (2.2)).*

We remark that the field is determined by the function F itself, not just by its zeroth level surface, so that it is defined in the entire space J^1.

Example 1. Suppose F is independent of y. The corresponding equation is called a *Hamilton–Jacobi equation*. The traditional notation for it is $F = H(x,p)$. A Hamilton–Jacobi equation is an equation of the form $H(x, \frac{\partial u}{\partial x}) = 0$. The equations of its characteristics have the form

$$x' = \frac{\partial H}{\partial p}\,, \quad p' = -\frac{\partial H}{\partial x}\,, \quad y' = p\,\frac{\partial H}{\partial p}\,. \tag{2.5}$$

The first two equations are the usual canonical Hamilton–Jacobi equations. (And this determines the choice of sign for the characteristic field indicated above: for the Hamiltonian of a free particle of unit mass $H = p^2/2$; with this choice of sign we obtain $x' = p$, that is, the momentum is numerically equal to the velocity.)

By projecting the system (2.5) onto the subspace with coordinates (x,p), we obtain a separate system of equations in this space. Thus we have carried out a factorization or splitting of the original equation (2.5). This vector field in the space (x,p) can also be invariantly defined, without resorting to coordinates. The surface of the original equation in J^1 was cylindrical, not depending on y. Such a surface projects very well onto the (x,p)-space. The Hamiltonian field is defined not only on the zeroth level surface of H, but everywhere. The function H usually has the physical meaning of energy.

Suppose, for example, $H = (p^2 - 1)/2$. Then the equation has the form $p^2 = 1$ or $\left(\frac{\partial u}{\partial x}\right)^2 = 1$, that is, it is the eikonal equation of geometric optics.

The system of equations of characteristics $x' = p$, $p' = 0$ describes the motion of particles along straight rays at constant speed (see Fig. 2.2).

If a Riemannian metric is defined on the manifold, the characteristics of the eikonal equation turn out to be the geodesics of this metric.

Fig. 2.2. The characteristics
of the eikonal equation

Example 2. Consider the Euler equation $u_t + uu_x = 0$. When $n = 2$, the space of 1-jets J^1 is five-dimensional, and $F(t, x, y, p_t, p_x) = p_t + yp_x$. Let us write the system of equations of the characteristics:

$$p'_t = -p_x p_t , \qquad t' = 1 ,$$
$$p'_x = -p_x^2 , \qquad x' = y , \qquad y' = p_t + p_x y = 0 .$$

The first two equations are the so-called conjugate equations of evolution of the derivatives, which we shall not study at the moment. Let us take up the remaining equations. The variable t represents time, and we write the remaining equations as $dx/dt = y$, $dy/dt = 0$, whence $d^2x/dt^2 = 0$, that is, the equation of the characteristics has turned out to be Newton's equation (see the previous lecture).

We now apply the characteristics to solve the equations in the general case.

Theorem 2. *Let Γ^n be the 1-graph of a solution, and suppose a characteristic passes through a point of this graph. Then the characteristic lies entirely in Γ^n, that is, the graph of a solution can be fibered into characteristics.*

PROOF. Consider a regular point z, and let ξ be the vector of the characteristic direction at the point z. The following are obvious:

1. $T_z\Gamma^n \subset T_zV^{2n}$, since Γ is a submanifold in V^{2n}.
2. $T_z\Gamma^n \subset K_z^{2n}$ by definition of K_z^{2n}, and therefore $T_z\Gamma^n \subset (T_zV^{2n}) \cap (K_z^{2n})$. We remark that the space $T_z\Gamma^n$ itself is n-dimensional.
3. All vectors in $T_z\Gamma^n$ are pairwise skew-orthogonal. Indeed, $\alpha|_{\Gamma^n} = 0$, since Γ^n is a 1-graph; therefore $d\alpha|_{\Gamma^n} = 0$, that is, $\omega^2|_{\Gamma^n} = 0$.

Exercise. Verify that in coordinate notation the vanishing of the form $\omega^2|_{\Gamma^n}$ follows from the symmetry of mixed partial derivatives.

Definition. A subspace of symplectic space all of whose vectors are pairwise skew-orthogonal is *isotropic* ("crazy" according to Lie).

For example, every line is isotropic on the symplectic plane.

Lemma. *In $2n$-dimensional symplectic space the dimension of an isotropic subspace is at most n.*

(Isotropic subspaces of maximal dimension exist and are called *Lagrangian* subspaces. An example of a Lagrangian subspace is a line in the symplectic plane.)

PROOF. Suppose an isotropic subspace is m-dimensional. Being isotropic, it is contained in its skew-orthogonal complement, and hence the dimension of the skew-orthogonal complement is at least m. We then have $2n - m \geq m$, so that $m \leq n$. The lemma is now proved. $\square$

We now return to the characteristic vector ξ at the point z.

Claim. *The vector ξ lies in $T_z \Gamma^n$.*

Suppose the contrary. Then, being characteristic, ξ is skew-orthogonal to $T_z V^{2n} \cap K_z^{2n}$. By virtue of item 2 above, ξ is skew-orthogonal to $T_z \Gamma^n$. But then the linear span of ξ and $T_z \Gamma^n$ is $(n+1)$-dimensional and is isotropic by virtue of item 3, all of which contradicts the lemma.

Thus, at each point the characteristic direction is tangent to the 1-graph of the solution, and so the entire characteristic lies on the graph. The theorem is now proved. $\square$

A recipe for constructing the 1-graphs of solutions now follows. (We remark that the existence of even one solution has not yet been proved.) One must take an $(n-1)$-dimensional submanifold of the surface V^{2n} that is not tangent to a characteristic and draw the characteristics through its point, thereby obtaining the 1-graph of a solution locally. To obtain a real graph it is necessary that the plane spanned by the tangent plane to an isotropic submanifold and the characteristic direction project one-to-one onto the x-space.

There is a standard method for constructing initial isotropic manifolds. Consider an $(n-1)$-dimensional surface γ^{n-1} in x-space, and suppose an initial function u_0 is defined on it. Substitute these data into the equation and regard it as an equation with respect to p. We have already defined the derivatives along the directions tangent to γ^{n-1}. The derivative in a transversal direction can be found from the equation using the implicit function theorem if F_p' is not tangent to the initial hypersurface γ^{n-1} at the point in question. Thus we obtain an $(n-1)$-dimensional isotropic initial submanifold; data of this type are called *Cauchy data*. For details see [1, § 8, pars. J and L].

Theorem 3. *Consider an initial $(n-1)$-dimensional submanifold $V^{n-1} \subset V^{2n}$ that is not tangent to a characteristic. Let the n-dimensional plane spanned by its tangent plane and the characteristic direction project isomorphically to x-space. Draw the characteristics through its points. Then we obtain the 1-graph of a solution locally (see Fig. 2.3).*

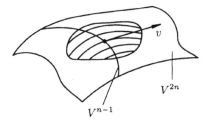

Fig. 2.3. Construction of the 1-graph of a solution

PROOF. The submanifold V^{n-1} is isotropic: $d\alpha|_{T_z V^{n-1}} = 0$. Drawing the characteristics through all of its points in a neighborhood of a regular point z, we obtain an n-dimensional submanifold Γ^n in V^{2n}. We shall prove that $d\alpha|_{\Gamma^n} = 0$. Let v be the vector field of characteristic directions. By a homotopy formula, $L_v\alpha = i_v d\alpha + d(i_v\alpha)$, where $i_v d\alpha(\xi) := d\alpha(v, \xi)$ and $i_v\alpha = \alpha(v)$. It is obvious that $i_v\alpha = 0$ and $i_v d\alpha(\xi) = d\alpha(v, \xi) = 0$ for every vector ξ that is tangent to V^{2n}, by definition of a characteristic direction. Thus, the derivative of the form in the direction of the field v on V^{2n} equals 0. Therefore the phase flow of the characteristic field maps the form into itself. This means that the value of the form on any vector equals the value of the same form on the vector translated back to the initial manifold.

It follows that $d\alpha$ is also mapped into itself by the field v. At the origin the space $T_z\Gamma^n$ is isotropic. (V^{n-1} is isotropic and v is characteristic.) This property carries over to any point z' of the manifold Γ^n by the phase flow of the field v: the space $T_{z'}\Gamma^n$ is isotropic. Thus $\alpha|_{\Gamma^n} = 0$. Hence Γ^n is an integral submanifold of the contact structure. Thus Γ^n is a Legendre submanifold. And by construction Γ^n is a submanifold of the equation. It remains to be proved that it is locally a 1-graph. That follows from the following general theorem.

Theorem 4. *Suppose an n-dimensional surface Γ^n in the space of 1-jets is a Legendre manifold (an integral manifold for the contact structure) and locally projects in a one-to-one manner onto the x-space. Then it is locally a 1-graph (see Fig. 2.4).*

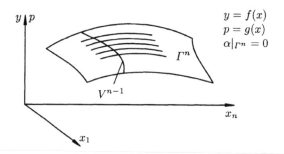

$$y = f(x)$$
$$p = g(x)$$
$$\alpha|_{\Gamma^n} = 0$$

Fig. 2.4. The integral surface as a 1-graph

PROOF. We write the condition for a Legendre manifold $\alpha\big|_{\Gamma^n} = 0$ (the surface is an integral surface for the contact structure):

$$\left(\sum_{i=1}^{n} \frac{\partial f}{\partial x_i}\, \mathrm{d}x_i - \sum_{i=1}^{n} g_i\, \mathrm{d}x_i = 0\right) \Longrightarrow \left(g_i = \frac{\partial f}{\partial x_i}\right),$$

since x_i are local coordinates on Γ^n. But this means that Γ^n is a 1-graph. □

Example. Let us solve the Cauchy problem for the Euler equation $u_t + uu_x = 0$. The initial curve in the (x, t)-plane is usually chosen to be the line $t = 0$. We prescribe an initial function $y = u_0(x)$, as in Fig. 2.5.

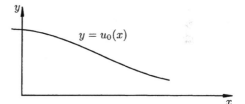

Fig. 2.5. The initial condition for the Euler equation

The equations of the characteristics are $t' = 1$, $y' = 0$, $x' = y$, and the transversality conditions are satisfied. We remark that for a quasi-linear equation everything can be considered in the space J^0, since the characteristics of a quasilinear equation in J^0 are the projections of the "true" characteristics from J^1, as shown in Fig. 2.6.

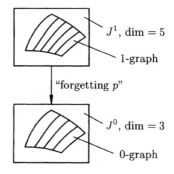

Fig. 2.6. The characteristics of a quasilinear equation in J^0 are projections of the characteristics of J^1

Thus we shall study the 0-graph in J^0. On the (x, y)-plane we draw a series of sections of the graphs for $t = \text{const}$, as in Fig. 2.7. The values of the solution at subsequent times are obtained by translation of the initial values along the characteristics.

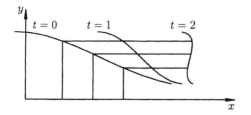

Fig. 2.7. The graphs
of the solution of the Euler
equation at successive times

We see that from some time on the curve ceases to be a graph. The integral surface of the equation ceases to project in a one-to-one manner onto the (x, t)-plane. The curve of critical values for the projection has a cusp, as in Fig. 2.8.

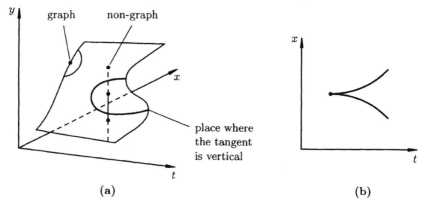

(a)   (b)

Fig. 2.8. (a) The integral surface ceasing to be a graph; (b) the curve of critical values of the projection

Physically the Euler equation describes the evolution of the velocity field of noninteracting particles. This model is good, for example, for describing star streams; the loss of uniqueness of the solution can be interpreted as the free passage of different streams through each other.

On the other hand, for large densities the particles begin to collide, and after a certain amount of time the Euler equation ceases to hold. It gets replaced by another equation that takes account of interaction, for example, Bürgers' equation $u_t + uu_x = \varepsilon u_{xx}$. For small ε its solutions approximate the solutions of the Euler equation up to the critical time, while for large values of time they are of shock wave type: the graph of u is nearly vertical in a small neighborhood (of order ε) of a moving point, as shown in Fig. 2.9.

To the right and left of this point the solutions are also close to solutions of Euler's equation. It is remarkable that Bürgers' equation can be solved explicitly – it reduces to the heat equation, which we shall study below.

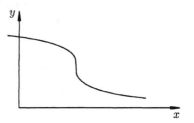

Fig. 2.9. Solutions of shock wave type

Remark. The 1-graph of a solution of a first-order equation consists of the characteristics of the hypersurface that defines the equation in the space of 1-jets, as in Fig. 2.10.

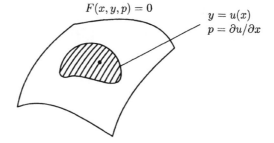

Fig. 2.10. The 1-graph of a solution in the space of 1-jets

Thus we have reduced the equation of the field to the equations of motion of particles. But, as Jacobi remarked, one can apply the theorem in the opposite direction: the equations of motion of particles can be reduced to the equation of the field. This is useful, since the systems of equations for particle motion cannot as a rule be solved explicitly.

If we find one solution of the equation of the field (a partial differential equation), we shall know a Legendre submanifold composed of characteristics. If we find another solution of the field equation, we shall have imposed another condition on the characteristics, and the integral manifold of intersection decreases in dimension by 1. If we succeed in finding sufficiently many solutions, we shall have reduced the dimension of the integral manifold. If a sufficiently large (finite-parameter) family of solutions of a partial differential equation are known, all the characteristics can be obtained by intersecting the corresponding Legendre manifolds until we isolate the (one-dimensional) characteristic itself.

This is Jacobi's method of integrating ordinary differential equations by representing them as the equations of the characteristics of some partial differential equation.

Jacobi applied this method to the solution of Hamiltonian systems – the systems of equations of characteristics for the Hamilton–Jacobi equation (in which F does not explicitly depend on y). One result was, for example, the famous theorem of Liouville on the integration of Hamiltonian systems for which a complete set of integrals in involution is known.

Astonishingly, this method could be "lifted" to integrate partial differential equations as well! To do this it is necessary to regard them as infinite-dimensional Hamiltonian systems. In this way the well-known sine–Gordon equation and the Korteweg–de Vries equation were integrated (see [2, Appendix 16]).

But it follows from our remark that the Jacobi method can be applied to a larger class of systems than Hamiltonian systems, namely to those that are the equations of the characteristics of some partial differential equation. To be sure, it is difficult to find out whether a specific vector field is characteristic for some equation. The corresponding theory seems to be undeveloped as yet; at least I do not know any applications of it to physically interesting problems.

Literature

1. Arnold, V.I.: Geometrical Methods in the Theory of Ordinary Differential Equations, 2nd edition. Springer, New York (1988)
2. Arnold, V.I.: Mathematical Methods of Classical Mechanics, 2nd edition. Springer, New York (1989)

Lecture 3

Huygens' Principle in the Theory of Wave Propagation

We consider space-time $M^{n+1} = B^n \times \mathbb{R}$, where B^n is "physical space," and $\mathbb{R}$ is the "time axis." (Most of what follows carries over word for word to the case of the bundle $M^{n+1} \to \mathbb{R}$ over the time axis, and a great deal of it also carries over to the case when the role of time is played by a fibration into "isochrones" or even the zeros of a nonclosed differential form "dt" on M.)

In geometric optics (and also in the calculus of variations and optimal control theory) a *cone of possible velocities of motion* is prescribed at each point of a manifold M.

Example. Let B be a Riemannian manifold. The graphs of all possible "motions" $q : \mathbb{R} \to B$ with unit velocity at each point is tangent to the quadratic cone (which defines the Riemannian metric) $\|dq\| = |dt|$. Thus, M is endowed with a field of Lorentz quadratic cones.

The tangent planes to the cone of possible directions at the point m that forms its vertex belong to the projectivization of the tangent space to M at that point: $PT_m^* M \approx \mathbb{R}P^n$.

It is often useful to endow these tangent planes with co-orientations (indicating the direction of propagation of a disturbance). The oriented tangent planes of the cone of possible velocities belong to the spherization $ST_m^* M$ of the cotangent space of space-time M. This spherization is diffeomorphic to the sphere S^n.

The non-co-oriented (resp. co-oriented) hyperplanes in the tangent space of a manifold are called the *contact elements* (resp. *co-oriented contact elements*) of the manifold. They form the manifold of the bundle of contact elements (resp. co-oriented contact elements) $PT^* M \to M$ (resp. $ST^* M \to M$) with fiber $\mathbb{R}P^n$ (resp. S^n), where $n = \dim M - 1$.

Thus the field of cones of possible velocities defines a hypersurface in the manifold of (possibly co-oriented) contact elements of space-time. In geometric optics this hypersurface is called the *Fresnel hypersurface*. It is a fundamental geometric object of geometric optics, calculus of variations and optimal control theory.

Remark. In the theory of wave propagation the vectors of the space dual to the velocity space are often called *retardations*.

The hypersurface constructed above can be described as the field of cones in the space T^*M of retardations on space-time.

On the manifolds of contact elements of any manifold (co-oriented or not) there is a remarkable geometric structure – the contact structure. That is the field of hyperplanes in the tangent spaces of the manifold of contact elements, which is invariantly determined by the bundle of the manifold of contact elements over the original manifold M.

Such a field exists and is unique; it is called the *tautological field* and is defined by the following construction. Each point of the space of the bundle of contact elements on M is a hypersurface in the tangent space of the original manifold M. The preimage of this plane under projection of the space of the bundle on M is the hyperplane of the tautological field at the initial point.

Remark. The tautological contact structure gives a condition on the velocity of motion of a contact element. This condition is called the *skate condition*. The meaning of the condition is that a skate (the contact element on the plane of a rink) can rotate freely in place and can move in the direction determined by itself, but resists any attempt to move it in transversal directions.

Let us consider the integral manifolds of the tautological contact structure.

Example 1. For any hypersurface in M the set of contact elements tangent to it forms the integral manifold of the tautological contact structure in PT^*M (or in ST^*M, if they are co-oriented). The dimension of this integral manifold is slightly less (by 1/2) than half the dimension of the manifold of contact elements.

Example 2. The set of contact elements at a point of the manifold M forms the integral manifold of the tautological contact structure in PT^*M (or ST^*M). The dimension of this integral manifold is the same as the dimension of the integral manifold of Example 1.

Example 3. For any submanifold (of any dimension) in M the set of contact elements on M tangent to it forms an integral submanifold of the tautological contact structure in PT^*M (or ST^*M) of the same dimension as in Examples 1 and 2.

Problem 1. Prove that the manifold of contact elements on M^{n+1} with its tautological structure has no integral submanifolds of dimension greater than n.

Problem 2. Do there exist smooth integral submanifolds of dimension n for the tautological contact structure on the manifold of contact elements in M^{n+1} not obtained by the construction of Example 3?

HINT. Consider the contact elements that are tangent to the semicubical parabola $x^2 = y^3$ on the plane ($n = 1$).

Definition. Integral manifolds of maximal dimension (which is n for a contact manifold of dimension $2n + 1$) are called *Legendre manifolds*.

Theorem 1 (the theory of support functions). *The manifold of 1-jets of functions from the sphere S^{n-1} into $\mathbb{R}^n$ (with its natural contact structure) is contact diffeomorphic to the manifold of co-oriented contact elements in $\mathbb{R}^n$ (with its tautological contact structure).*

PROOF. We denote by $q \in S^{n-1} \subset \mathbb{R}^n$ a point of the unit sphere in Euclidean space. The tangent vectors to the sphere at the point q can be regarded as the vectors in the space orthogonal to q. They can also be regarded as cotangent vectors, since the Euclidean structure identifies a vector p with the linear functional $(p, \cdot)$. Thus the 1-jet of a function f on S^{n-1} is given by a triple $\left(q \in S^{n-1},\ p = df\big|_q \in \mathbb{R}^n,\ z = f(q)\right)$.

The required diffeomorphism is defined by the formula

$$Q = q + pz\,,\quad P = q\,,$$

where Q is the point of application of the contact element of $\mathbb{R}^n$ co-oriented by the vector P normal to it. □

Problem. Prove that: 1) this is indeed a diffeomorphism $J^1(S^{n-1}, \mathbb{R}) \approx ST^*\mathbb{R}^n$ ($\approx S^{n-1} \times \mathbb{R}^n$); 2) it maps the natural contact structure in J^1 into the tautological contact structure ST^*.

HINT. Consider what the Legendre manifolds map to.

Remark. The distance from the origin to the tangent plane orthogonal to q of a convex hypersurface in $\mathbb{R}^n$ is called the *value of the support function* of the given hypersurface at the point $q \in S^{n-1}$. The support function is defined on the sphere of (outward) normals and defines a hypersurface.

A function defined on the sphere determines a hypersurface in the space (the envelope of the family of hypersurfaces corresponding to the function). This hypersurface, however, may have singularities.

Problem. Study the curves defined by the functions $x^2 + 2y^2 + t$ (where t is a parameter) on the circle $x^2 + y^2 = 1$.

Let us return to the Fresnel hypersurface of retardations of fronts lying in the manifold ST^*M^{n+1} of co-oriented contact elements of space-time.

Definition. The *rays* (on this hypersurface) are its characteristics, that is, the integral curves of the field of characteristic directions of the hypersurface.

We recall that a characteristic direction of a hypersurface at a point of a manifold with a contact structure (defined as the field of zeros of a differential 1-form α) is the skew-orthogonal complement (in the sense of the symplectic structure $d\alpha$ on the plane $\alpha = 0$ at each point) to the intersection of the tangent hypersurface of the hypersurface with the hyperplane of the contact structure.

Example. Let $(q_1, \ldots, q_n)$ be the coordinates in "physical space" B^n, let $q_0 = t$ be time, and let $(p_0, p_1, \ldots, p_n)$ be the corresponding components of the momentum. The tautological contact structure is given locally by the 1-form $\alpha = p_0 \, dq_0 + p_1 \, dq_1 + \cdots + p_n \, dq_n$, where $[p_0 : p_1 : \cdots : p_n]$ are homogeneous coordinates in $\mathbb{R}P^n$ and $q_0 = t$. In affine coordinates, for which $p_0 = -1$, we have $\alpha = -dt + p_1 \, dq_1 + \cdots + p_n \, dq_n$. The retardation hypersurface is given by the equation

$$p_0 = H(p_1, \ldots, p_n; q_1, \ldots, q_n, t) \, ,$$

where H is a homogeneous function of degree 1 with respect to the variables p_i. The equation of the characteristics has the form

$$\frac{dp_i}{dt} = -\frac{\partial H}{\partial q_i} \, , \quad \frac{dq_i}{dt} = \frac{\partial H}{\partial p_i} \quad (i = 1, \ldots, n) \, .$$

Thus the rays are defined by the Hamilton equations, the Hamiltonian being given by the retardation hypersurface.

Let us now consider the geometry of wave propagation in a medium in which the local velocity of propagation of disturbances is defined by this retardation hypersurface in the space of contact elements of space-time.

A typical example is the family of equidistants of a submanifold of a Riemannian manifold. The retardation hypersurface defines a second-order hypersurface in each projectivized cotangent space of space-time. The family of equidistants (hypersurfaces lying at distance t from a given initial hypersurface in physical space) can be regarded as a single hypersurface in space-time (whose sections by different isochrones $t = \text{const}$ give the entire family of equidistants). Similarly one can define the *large front* that describes the propagation of fronts of perturbations by means of a single hypersurface in space-time with a more general retardation hypersurface in the manifold of contact elements of space-time.

Let us consider the contact elements of space-time that are tangent to the large front. They all belong to the retardation hypersurface (that is the content of the local law of propagation of disturbances). The following statement is obvious.

Proposition. *The contact elements tangent to a large front form a Legendre submanifold of the tautological contact structure of the space of contact elements of space-time that projects to the large front and lies in the retardation hypersurface.*

Let us now consider the initial condition given by the instantaneous front of a disturbance at time $t = 0$. This front determines a submanifold of codimension 1 on the large front, and this submanifold is also integral for the tautological contact structure of the manifold of contact elements of space-time. It is not a Legendre manifold, however, since its dimension is one less than that of the Legendre manifolds.

This initial integral submanifold consists of contact elements of space-time that are tangent to the initial front and in addition belong to the retardation hypersurface. It is this last condition that renders it possible (under suitable nondegeneracy conditions) to choose the contact element of space-time that is tangent to the cone of possible velocities out of the one-parameter family of contact elements of space-time containing the tangent space to the initial front at the given point.

The basis of the entire theory of propagation of disturbances is provided by the following simple and general fact of contact geometry (discovered in essence by Huygens and hence deserving the name "Huygens' principle").

Theorem 2. *A Legendre submanifold of a hypersurface in contact space contains, along with each point, the entire characteristic of that surface passing through the point.*

It is assumed here that the tangent plane to the hypersurface nowhere coincides with the contact plane $\alpha = 0$.

PROOF. Along a Legendre manifold we have $\alpha = 0$, so that $d\alpha = 0$ in the tangent plane to the manifold. If a characteristic direction vector ξ did not lie in this plane, it would be skew-orthogonal to it in the sense of the form $d\alpha$, and, adjoining ξ to the tangent plane of the Legendre manifold, we would obtain a plane skew-orthogonal to itself in the symplectic vector space $\alpha = 0$ having dimension greater than half the dimension of the space. This is impossible, and therefore the characteristic directions are tangent to the Legendre manifold, as asserted. $\square$

Corollary. *The Legendre manifold corresponding to a large front is obtained from the initial condition by the following construction. The initial front is lifted to the integral manifold of the space of contact elements of space-time lying in the retardation hypersurface. We then consider the characteristics of this hypersurface passing through points of the integral manifold just constructed. They form a Legendre manifold. The projection of this manifold on space-time is a large front. The sections of the large front by the isochrones $t = $ const are the instantaneous fronts.*

This constitutes a description of the propagation of disturbances using waves (fronts) and rays (characteristics).

Since the characteristics are determined by ordinary differential equations, each of them is determined by any one of its points. They will be the same under completely different initial conditions for which only the tangent hyperplanes of the initial fronts coincide at one point at the initial time. Physically this means that infinitely small pieces of a wave front propagate (along the characteristics) independently of one another. That makes it possible to invoke particles (whose motion is determined by ordinary differential equations, namely the Hamilton equations for the characteristics of the retardation hypersurfaces) to describe the propagation of the front (waves, which are determined by partial differential equations).

In particular, the initial condition can even be taken to be a point front (corresponding to the Legendre manifold forming a fiber of the bundle of contact elements). The theorem is applicable in this case also (although it appears to be degenerate from the point of view of the corollary just given).

A "spherical" wave corresponds to a point initial front. Comparing the propagation of disturbances from an arbitrary initial front with those from one of its component point sources, we see that the Legendre manifolds corresponding to it contain a common characteristic. It follows that the instantaneous front is subsequently tangent to the spherical front of the original point source at a point on the ray emanating from the source in the direction determined by the direction of the initial front.

It follows that at time t the instantaneous front is the envelope of the family of spherical fronts of point sources belonging to the original front. This was the original formulation of Huygens' principle (known under different names in different branches of mathematics: Pontryagin's principle, the canonical Hamilton equations, and so forth).

Incidentally, the equations of characteristics are precisely the infinitesimal version of Huygens' assertion about the envelope, corresponding to infinitely small values of t and consequently to infinitely small spherical fronts.

Naturally it was assumed in the geometric theory described above that the necessary determinants are nonzero, so that the implicit function theorem could be applied. For example, the field of Lorentz cones in space-time cannot be tangent to the isochrones $t = $ const at any point, and so on. It is not difficult to verify that these conditions hold automatically in such examples as the problems of Riemannian geometry, even with a time-dependent metric. In other problems, for example, in optimal control theory, the situation often turns out to be more complicated and the need to investigate singularities arising when the determinant vanishes constitutes a major difficulty.

The fact that the characteristics passing through points of the initial integral manifold form a Legendre manifold needs the proof. We shall not go into detail on this point, since it is exactly the same as the proof of Theorem 3 given in the preceding lecture.

Lecture 4

The Vibrating String (d'Alembert's Method)

We consider a string, which can be pictured as the actual string of a musical instrument, attached to a sounding board, as in Fig. 4.1.

Fig. 4.1. The vibrating string

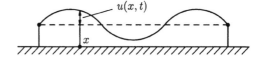

Let $u(x,t)$ be the displacement of the string at the point x at time t in the orthogonal direction from its equilibrium position. It is known that if the string is homogeneous and the displacements are small compared with the length of the string, the function $u(x,t)$ satisfies the following linear second-order partial differential equation:

$$u_{tt} = a^2 u_{xx} \ . \tag{4.1}$$

We shall not discuss the derivation of this equation in the present lecture. Here a is a constant having the physical dimension of velocity, as will be shown below.

4.1. The General Solution

Problem 1. Show that the change of variables

$$\xi = x - at \ , \quad \eta = x + at$$

reduces the equation (4.1) to the form

$$u_{\xi\eta} = 0 \ . \tag{4.2}$$

(When $a = 1$, this change of variables is the composition of a rotation and a dilation in the plane of the independent variables.)

The equation (4.2) is easy to solve. We write it in the form $(u_\xi)_\eta = 0$, which means that the function u_ξ is independent of η, that is, constant along vertical lines on the (ξ, η)-plane, as shown in Fig. 4.2.

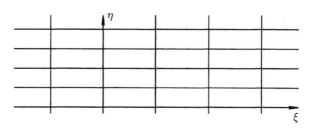

Fig. 4.2. The characteristics of the vibrating string equation

Hence $u_\xi = f(\xi)$, that is, the differential equation has been reduced to a simple equation of the form $F' = f$, so that $u = \int f(\xi)\,d\xi = F(\xi) + G(\eta)$; we remark that the constant of integration in general depends on the line along which the integration occurs, that is, it depends on η.

If we seek a solution in the class of functions that are continuous at every stage of the computation, we obtain a general solution of the form $f(\xi) + g(\eta)$, where the function f is smooth and g is continuous. We have thus arrived at a "solution" that in general might have no derivatives! Moreover, if we solve in the opposite order, we obtain an asymmetry in regard to smoothness. However, equality of mixed partial derivatives holds only in the case of sufficiently smooth functions. This observation serves as motivation for the introduction of generalized functions (distributions), a class in which one can consider a solution of the form $f(\xi) + g(\eta)$ with nonsmooth f and g.

In the present lecture we assume that there exists a solution in the class of sufficiently smooth functions, so that f and g are sufficiently smooth.

Thus, $u(x, t) = f(x - at) + g(x + at)$ is the general solution of the equation (4.2).

4.2. Boundary-Value Problems and the Cauchy Problem

The Cauchy problem reveals both the similarities and the differences between the theories of ordinary and partial differential equations. In the theory of ordinary differential equations the phase space is finite-dimensional, whereas in this book we shall be dealing with an infinite-dimensional phase space.

The Cauchy problem for the vibrating string consists of the equation (4.1) with initial conditions

$$u\big|_{t=0} = \varphi(x)\,, \quad u_t\big|_{t=0} = \psi(x)\,. \tag{4.3}$$

The value of the second derivative at the initial instant of time need not be prescribed, since it can be determined from the equation.

We assume that $x \in \mathbb{R}$, that is, the string is infinite. This model gives a good approximation to physical reality, provided we consider deviations that are small compared to the length of the string over brief time intervals.

The first boundary-value problem includes the equation (4.1) for $x \in (0, l)$, the initial conditions (4.3), and the boundary conditions

$$u\big|_{x=0} = u\big|_{x=l} = 0 \, , \tag{4.4}$$

which express the fact that the finite string is clamped at both ends.

This problem also describes the longitudinal vibrations of a rod with fixed ends. In this case $u(x, t)$ is the displacement of the point x of the rod from its equilibrium position at time t, as in Fig. 4.3.

Fig. 4.3. The vibrating rod

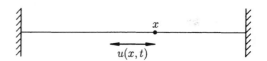

One can consider the vibrations of a rod with one or both ends free, expressed by the condition

$$u_x\big|_{x=0} = 0 \quad \text{and/or} \quad u_x\big|_{x=l} = 0 \, . \tag{4.5}$$

The second and third boundary-value problems are obtained as combinations of these conditions.

The fourth boundary-value problem involves consideration of the periodic boundary conditions

$$u(x, t) = u(x + l, t) \, , \tag{4.6}$$

so that the function u is defined on a circle.

We are now going to obtain the complete solutions of all these problems (some of them in the exercises).

4.3. The Cauchy Problem for an Infinite String. d'Alembert's Formula

Thus, consider the Cauchy problem $u_{tt} = a^2 u_{xx}$, $u\big|_{t=0} = \varphi(x)$, $u_t\big|_{t=0} = \psi(x)$.

Theorem (d'Alembert's formula). *The solution of the Cauchy problem is given by the formula*

$$u(x, t) = \frac{\varphi(x - at) + \varphi(x + at)}{2} + \frac{1}{2a} \int\limits_{x-at}^{x+at} \psi(y) \, dy \, . \tag{4.7}$$

OUTLINE OF THE PROOF. We know that the general solution of the equation
has the form $u(x,t) = f(x - at) + g(x + at)$. Consider the term $f(x - at)$. At
a fixed time t the graph of the function $f(x - at)$ is the graph of the function
$f(x)$ shifted to the right by at when $t > 0$. This term is called the *forward
wave*. The analogous term $g(x + at)$ is called the *reverse wave*. We first put
$t = 0$ in the formula for the general solution; then, differentiating with respect
to t in this same formula, we obtain the following system:

$$\begin{cases} u\big|_{t=0} = \varphi(x) = f(x) + g(x) , \\ u_t\big|_{t=0} = \psi(x) = -af'(x) + ag'(x) . \end{cases} \tag{4.8}$$

Solving this system, we find f and g, substitute them into the formula for the
general solution, and obtain (4.7). $\square$

Problem 2. Fill in the details of this proof.

"Movie" Problems

Using d'Alembert's formula, one can draw the successive "frames of a movie
of the vibrations of the string," knowing the graphs of the Cauchy conditions.

Example. Let $\psi \equiv 0$, and let the graph of φ have the form shown in Fig. 4.4,
where $\varphi \neq 0$ on an interval of length 1.

, $\psi \equiv 0$ **Fig. 4.4.** The initial conditions

Then by d'Alembert's formula, which has the form $u(x,t) = \big(\varphi(x - at) + \varphi(x+at)\big)/2$ in the present case, we obtain the shape of the string at successive
times, as in Fig. 4.5.

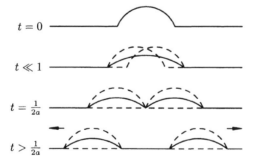

Fig. 4.5. The shape
of the string at
successive
times

Problem 3. Draw the movie for the initial conditions $\varphi \equiv 0$ and ψ a function
of the form ⌒ .

4.4. The Semi-Infinite String

This problem includes the equation (4.1), the initial conditions (4.3), and the boundary conditions $u\big|_{x=0} \equiv 0$ (fixed endpoint) or $u_x\big|_{x=0} \equiv 0$ (free endpoint).

By combining the boundary conditions and the form of the initial data we obtain the following table of problems:

initial conditions	$\varphi:$ ⌒ $\psi \equiv 0$	$\varphi \equiv 0$ $\psi:$ ⌒
boundary conditions		
free endpoint at $x = 0$	4	5
fixed endpoint at $x = 0$	6	7

Problems 4–7. Draw the movies of the motion of the string for the conditions shown in the table.

HINT. The problem for the semi-infinite string can be reduced to the problem for the infinite string so that the solution of the problem for the infinite string, when restricted to a half-line, yields the solution of the problem for the semi-infinite string. To do this we must extend the initial conditions to the entire line in such a way that the solution satisfies the boundary conditions at the point $x = 0$.

Parity considerations are helpful in doing this. The initial condition can be extended to the entire line as an even function provided the relation $u_x\big|_{x=0} = 0$ holds at the initial instant of time. But will the solution be an even function of x at all times? First of all, this can be deduced from d'Alembert's formula; second, one can make use of the following remarkable idea.

The uniqueness of the solution of the Cauchy problem was proved in the derivation of d'Alembert's formula. This equation is invariant under the transformation $x \mapsto -x$. If the initial condition is even, that is, also invariant under this transformation, we then have two solutions $u(x,t)$ and $u(-x,t)$. Since the solution is unique, these two solutions coincide, so that $u(x,t) = u(-x,t)$ and the solution is an even function.

This is a general idea: If the problem supports some symmetry and the solution is unique, then the solution must also support that symmetry.

Similarly one can use an odd extension when the condition $u\big|_{x=0} = 0$ holds.

4.5. The Finite String. Resonance

The method of d'Alembert is not very convenient for solving the boundary-value problems in the case of a finite string. Later on we shall develop another very powerful method to handle this case. At present we illustrate the application of d'Alembert's method to the problem of forced vibrations of a finite string. This problem consists of the equation (4.1), the initial conditions (4.3), and the following boundary conditions: $u\big|_{x=0} = f(t)$, $u\big|_{x=l} = 0$.

Problem 8. Find the general solution of this problem.

HINT. The values brought from the boundary and the initial interval $t = 0$ along the characteristics $x - at = $ const, $x + at = $ const contribute to the solution at the point (x, t). The characteristics undergo bending as they are reflected from the boundary. As a result, the value of the solution is an alternating sum of the values at the nodes of the resulting broken lines, as shown in Fig. 4.6. The initial conditions can be arbitrary; for simplicity you may first take them to be zero.

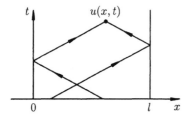

Fig. 4.6. How to obtain the value of the solution at a given point

Problem 9. Suppose the function $f(t)$ is periodic with period T. Will there exist periodic solutions with periodic boundary conditions? If so, will the other solutions be periodic?

ANSWER. There is no periodic solution, for example, in cases when resonance occurs (when the period of the external forces T is commensurable with the "periods of natural oscillation" $2l/a$). For almost all (in the Lebesgue sense) values of T there exists a periodic solution (at least if the function is smooth). This follows from number-theoretic considerations (the metric theory of Diophantine approximations).

   In conclusion we remark that all these results have required only elementary mathematical analysis, remaining within the scope of the theory of differentiation and integration. The following sections require more powerful tools of analysis.

4.6. The Fourier Method

Let us return to the equation (4.1): $u_{tt} = a^2 u_{xx}$. We can regard the right-hand side as a differential operator that maps one function space into another (or into itself, if we consider the space of infinitely smooth or analytic functions).
   We write the equation (and any problem we may consider along with it) more abstractly, for example:

$$u_{tt} = Au, \qquad u\big|_{x=0} = u\big|_{x=l} = 0 \quad \text{or} \quad u(x + 2\pi, t) = u(x, t) .$$

We shall not specify any initial conditions, since we wish to learn how to solve the problem for arbitrary initial conditions.

Thus A is a linear operator. A point of the phase space is a function u satisfying our boundary conditions.

The operator A has certain supplementary properties in addition to linearity. Let us see what these properties yield, for simplicity starting with finite-dimensional models.

Model 1. We consider the linear ordinary differential equation $u_t = Au$, where u is a vector in Euclidean space $\mathbb{R}^N$ and A is a self-adjoint operator: $(Ax, y) = (x, Ay)$, the parentheses denoting the inner product. Then the operator A has an orthogonal basis consisting of eigenvectors X_k, and the corresponding eigenvalues λ_k are real. The functions $\mathrm{e}^{\lambda_k t} X_k$ form a fundamental system of solutions. The general solution has the form $u(t) = \sum c_k \mathrm{e}^{\lambda_k t} X_k$. It is easy to find a solution with prescribed initial condition $u(0) = \varphi$ (in this case these are a finite-dimensional vectors). We substitute $t = 0$: $\varphi = \sum c_k X_k$, and we need to find c_k. We then take the inner product with X_j, use the orthogonality of the basis, and obtain the result that $c_j = (\varphi, X_j)/(X_j, X_j)$.

The similarity to the wave equation here is that our second-order differential operator $\mathrm{d}^2/\mathrm{d}x^2$ is self-adjoint in a suitable space.

Differential Operators in the Space of Trigonometric Polynomials

One of Hilbert's most remarkable ideas was to regard function spaces as Euclidean spaces. This idea is the foundation of functional analysis.

But at present we shall restrict ourselves to finite-dimensional spaces of trigonometric polynomials. For any nonnegative integer N we consider the space

$$E_N = \left\{ u(x) = \sum_{k=-N}^{k=N} c_k \mathrm{e}^{ikx}, \ x \in S^1, \ c_k \in \mathbb{C} \right\}.$$

As long as the summation extends only from $-N$ to N, we remain within the limits of linear algebra. If it extends from $-\infty$ to $+\infty$, we go into the area of Fourier series, which is a branch of functional analysis.

On the space E_N we introduce a Hermitian form:

$$(f, g) := \int_0^{2\pi} f\bar{g} \, \mathrm{d}x \,.$$

Problem 10. Verify that all the axioms of a space with a Hermitian inner product are satisfied.

The norm of a polynomial is defined by the formula $\|f\|^2 = (f, f)$.

Problem 11. Prove that the polynomials $e_k = e^{ikx}$, $k \in \{-N, \ldots, N\}$, form an orthogonal basis.

SOLUTION. $\int_0^{2\pi} e^{ikx} e^{-ilx} \, dx = 2\pi$ if $k = l$. But if $k - l = m \neq 0$, then the integral equals $\int_0^{2\pi} e^{imx} \, dx = \frac{1}{im} e^{imx} \big|_0^{2\pi} = 0$.

Model 2. We consider a periodic boundary-value problem for the wave equation with initial conditions in the space E_N:

$$u_{tt} = a^2 u_{xx} \,, \quad u(x + 2\pi, t) = u(x, t) \,,$$
$$u\big|_{t=0} = \varphi \,, \quad u_t\big|_{t=0} = \psi \,, \quad \varphi, \psi \in E_N \,.$$

In the next lecture we shall reduce this problem to a system of ordinary differential equations of the form $u_{tt} = Au$, that is, a system analogous to that of Model 1.

Lecture 5

The Fourier Method (for the Vibrating String)

Consider the periodic boundary-value problem for the vibrating string:

$$u_{tt} = a^2 u_{xx} , \quad x \in S^1 = \mathbb{R}/2\pi\mathbb{Z} , \tag{5.1}$$

$$u\big|_{t=0} = \varphi(x) , \quad u_t\big|_{t=0} = \psi(x) . \tag{5.2}$$

The problem can be regarded as a Cauchy problem on the circle S^1. On a connected compact manifold with boundary (for example, on an interval) it would be necessary to prescribe some boundary conditions.

5.1. Solution of the Problem in the Space of Trigonometric Polynomials

Consider the space of complex-valued trigonometric polynomials

$$E_N = \left\{ \sum_{-N}^{N} a_k e^{ikx} , \ a_k \in \mathbb{C} \right\} .$$

(One can also consider real-valued polynomials; the necessary and sufficient condition for the polynomial to be real-valued is $a_k = \overline{a_{-k}}$.)

Suppose the initial data φ and ψ belong to E_N. Then the solution $u(\cdot, t)$ also belongs to E_N. There is a Hermitian product in E_N:

$$(f, g) = \int_{S^1} f(x)\overline{g(x)}\, dx , \quad \|f\|^2 = (f, f) .$$

The following properties are easy to verify:

1. The monomials $\{e^{ikx}\}$ form an orthogonal basis of the space. This basis is not normalized: $\|e^{ikx}\| = \sqrt{2\pi}$. We introduce the notation $X_k = e^{ikx}$.
2. The operator $L = a^2 \frac{d^2}{dx^2}$ mapping E_N into itself is self-adjoint, having the X_k as eigenvectors:

$$LX_k = -a^2 k^2 X_k .$$

5.2. A Digression

Consider the Euclidean space $\mathbb{R}^n$ and a self-adjoint nonsingular operator L : $\mathbb{R}^n \to \mathbb{R}^n$. Let $\{X_k\}$ be an orthogonal basis of eigenvectors and $\lambda_k = -\omega_k^2$ the eigenvalues.

Consider the problem $\ddot{x} = Lx$, $x(0) = \varphi$, $\dot{x}(0) = \psi$.

The vector-valued functions $\sin(\omega_k t) X_k$ and $\cos(\omega_k t) X_k$ form a fundamental system of solutions of this problem. Indeed, the dimension of the solution space of the problem is $2n$ (as can be proved by reducing the problem to a system of first-order equations $\dot{x} = p$, $\dot{p} = Lx$). These $2n$ functions are solutions and, in addition, are linearly independent (verify this!).

Hence the general solution has the form

$$x(t) = \sum \left(a_k \cos(\omega_k t) X_k + b_k \sin(\omega_k t) X_k \right).$$

It is also easy to find a solution with prescribed initial conditions:

$$x(0) = \sum a_k X_k \implies a_k = \frac{(\varphi, X_k)}{(X_k, X_k)},$$

$$\dot{x}(0) = \sum b_k \omega_k X_k \implies b_k = \frac{(\psi, X_k)}{\omega_k (X_k, X_k)}.$$

5.3. Formulas for Solving the Problem of Section 5.1

In this case we write u instead of x, we have E_N instead of $\mathbb{R}^n$, and $L = a^2 \frac{\partial^2}{\partial x^2}$. For simplicity we assume $a = 1$. We have $X_k = e^{ikx}$, $\omega_k = k$.

Then the general solution of the Cauchy problem has the following form:

$$u(x,t) = \sum \left(a_k \cos(kt) + b_k \sin(kt) \right) e^{ikx}, \tag{5.3}$$

$$a_k = \frac{1}{2\pi} \left(\varphi, e^{ikx} \right), \tag{5.4}$$

$$b_k = \frac{1}{2\pi k} \left(\psi, e^{ikx} \right), \quad k \neq 0. \tag{5.5}$$

Here the summation on k extends from $-N$ to N.

Thus, the process of solving the Cauchy problem in the class E_N requires nothing more than linear algebra. But it is noteworthy that the solution, even in the general case, is given by formulas (5.3)–(5.5) (with summation from $-\infty$ to $+\infty$).

5.4. The General Case

Let φ and ψ belong to $C^\infty(S^1)$. (One can also consider the case of analytic initial data or, in the other direction, data with finite smoothness.)

Theorem. *The solution of the Cauchy problem* (5.1), (5.2) *is given by formulas* (5.3)–(5.5) *with summation from* $-\infty$ *to* $+\infty$.

To prove this result one must verify that the series converges, that it can be differentiated termwise, and that the initial conditions are satisfied.

5.5. Fourier Series

We need to prove that the series (5.3) converges along with all its partial derivatives up to order two inclusive.

Consider the space $C^\infty(S^1)$ with the Hermitian product

$$(f,g) = \int\limits_{S^1} f\bar{g}\,dx\,, \quad \|f\|^2 = (f,f)\,.$$

This space is not complete in this norm, and so it is called a *pre-Hilbert space*. (A complete space with a Hermitian product is called a Hilbert space.) In this space the system of functions $\{X_k\} = \{e^{ikx}\}$ forms a "basis" in a certain sense.

We remark that the series (5.3) can be rewritten as a series of exponentials by using Euler's formulas. The result is a series of terms of the form $e^{ik(t+x)}$ and $e^{ik(t-x)}$. Such a series is easy to differentiate.

To prove that the series (5.3) converges together with all the series obtained from it by termwise differentiation, it suffices to prove that the coefficients a_k and b_k tend to 0 faster than any power of k.

To prove that the initial conditions are satisfied, it is necessary to show that the Fourier series of a function converges to that function. It is in this sense that the system $\{X_k\}$ is a basis:

If $\quad \varphi \sim \sum a_k X_k\,, \quad a_k = (\varphi, X_k)/(X_k, X_k)\,, \quad$ then $\quad \sum a_k X_k \to \varphi\,.$

We remark that the part of the proof in which the mere convergence of the Fourier series of the initial data is proved (without regard to the limiting function) does not differ from the proof of the convergence of the series (5.3). To do this, as noted above, it suffices to establish that the $|a_k|$ decrease rapidly.

5.6. Convergence of Fourier Series

Lemma 1. *The Fourier coefficients of a function in* $C^\infty(S^1)$ *decrease faster than any power of the exponent.*

PROOF. The following estimate is obvious:

$$|a_k| \leq \left(\max_{S^1} |\varphi|\right) \cdot 2\pi\,,$$

that is, the coefficients are at least bounded. We now integrate by parts:

$$a_k = \frac{1}{2\pi} \int \varphi(x)\, e^{-ikx}\, dx$$

$$= \frac{1}{-2\pi ik} \int \varphi(x)\, d(e^{-ikx})$$

$$= \frac{1}{-2\pi ik} \left(\varphi(x)\, e^{-ikx} \big|_0^{2\pi} - \int \varphi'(x)\, e^{-ikx}\, dx \right)$$

$$= \frac{1}{2\pi ik} \int \varphi'(x)\, e^{-ikx}\, dx .$$

For the same reason as at the beginning of the proof, the last integral is bounded uniformly with respect to k. Continuing, one can again integrate by parts as many times as necessary.

As a result we find that $\forall m\ \exists C_{m,\varphi} : \left|(\varphi, e^{ikx})\right| < C_{m,\varphi}|k|^{-m}$. The lemma is now proved. □

Remark. For a function of class C^m, we find that the Fourier coefficients decrease like $1/|k|^m$.

Problem. Suppose the function φ belongs to the class $C^\omega(S^1)$, that is, φ is holomorphic in the strip $|\operatorname{Im} z| \leq \beta$ and periodic: $\varphi(z + 2\pi) \equiv \varphi(z)$. Prove that then the Fourier coefficients decrease exponentially:

$$|a_k| < C e^{-\beta|k|} .$$

HINT. Shift the path of integration in the integral that defines the Fourier coefficients by $\pm i\beta$ (depending on the sign of k).

Is the converse true? That is, if this last estimate holds for the Fourier coefficients, does the function admit a holomorphic extension to the strip?

ANSWER. The sum of the series is holomorphic *in the interior* of the strip.

Remark. If a function is continuous, its Fourier series converges to it in the metric of L^2. The contemporary Swedish mathematician Carleson has shown that pointwise convergence holds on a set of full measure.

Lemma 2. *The Fourier series of a function in C^2 converges to the function.*

PROOF. The Fourier series of a function φ in C^2 converges uniformly to some function $\psi \in C^0$ (by Remark after Lemma 1). We must show that $\varphi \equiv \psi$. Suppose not. Then $\delta(x) := \varphi(x) - \psi(x) \not\equiv 0$. Nevertheless δ has a zero Fourier series: $\forall k\ (\delta, X_k) = 0$.

Let us attempt to approximate δ with a trigonometric polynomial:

$$\left\| \delta - \sum c_k X_k \right\|^2 = \|\delta\|^2 + \left\| \sum c_k X_k \right\|^2 \geq \|\delta\|^2 > 0 . \qquad (*)$$

We see that it is impossible to get an arbitrarily close approximation. However, by the Weierstrass theorem, one can always approximate a continuous function δ uniformly to any degree of precision using trigonometric polynomials. That contradicts $(*)$. The lemma is now proved. $\square$

Remark. One can consider the Fourier method for the problem on an interval with various boundary conditions. But we have actually developed a method of solving the completely general problem $u_{tt} = Au$, where u is a function on an arbitrary manifold. One need only find the eigenfunctions and eigenvalues for the operator A on the manifold, construct an orthogonal basis of eigenfunctions, and then follow the outline above.

For the Laplacian $A = \Delta$ on, for example, a manifold with a Riemannian metric, a rich theory thereby results. It forms part of the spectral theory of differential operators.

5.7. Gibbs' Phenomenon

The Fourier series of a discontinuous periodic function cannot converge uniformly to it, but it can converge pointwise. For example, such is the case for discontinuous piecewise-smooth functions: the sum of the series coincides with the function on the intervals of smoothness. The sequence of *graphs* of the partial sums of the Fourier series of a piecewise-smooth function converges (uniformly!), but not to the graph of the original function. Rather it converges to a different curve. This curve is obtained from the graph of the given discontinuous function by adjoining vertical line segments above the points of discontinuity. It is interesting that *these segments are longer than the segments joining the parts of the graph to the right and left of the point of discontinuity.* Here the lengths of the additional vertical tails extending above and below the graph of the original function always constitute the same portion (each approximately 9%) of the magnitude of the discontinuity, as in Fig. 5.1.

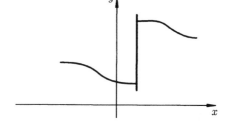

Fig. 5.1. Gibbs' phenomenon

To prove this amazing fact (known as *Gibbs' phenomenon*) it suffices to carry out the computation of the maximum and minimum of the partial sum of the Fourier series of a very simple discontinuous function of period 2π (say, the function equal to x for $|x| < \pi$). The general case can be reduced to this one by removing the discontinuities through subtraction of a suitable linear combination of translates of this very simple discontinuous function.

Gibbs' phenomenon leads to some interesting artefacts in tomography (where the function of two variables is, for example, the X-ray density of a planar section of a human body, reconstructed by a computer using the summation of a (double) Fourier series).[1]

Since the density is a discontinuous function (for example, because of the presence of bones), Gibbs' phenomenon arises. It manifests itself here in the form of additional lines to the lines of discontinuity: the double tangents and the inflection tangents of real boundaries of tissues of differing density (think about why this is so).

---

[1] The coefficients of this series are the Fourier coefficients of the so-called *Radon transform* of the original density. The Radon transform is the integral of the original density along a line, regarded as a function of the line; here it is regarded as a function of one variable, namely the distance of a line of a given pencil of parallel lines from a single one of these lines.

Lecture 6

The Theory of Oscillations. The Variational Principle

It is an experimental fact well-known to physicists that the laws of nature can be described by variational principles. The absence of rational grounds for this fact has generated attempts to interpret it theologically, philosophically, and in other ways, as, for example, in the writings of Voltaire, Maupertuis and others in the collection *Variational Principles of Mechanics*.

The variational principle asserts that "nature acts in the shortest way." For example, according to Fermat's principle, rays of light propagate along the shortest path. The principle of least action was stated by Hamilton in a form close to the modern one:

$$\delta \int L \, \mathrm{d}t = 0 .$$

The integral that occurs in this formula is called the *action integral*. The principle describes the motion of some mechanical system. The Lagrangian $L(q, \dot{q}, t)$ is a function of the state and velocity of the motion in the corresponding configuration space. In this form the principle reads as follows: "A motion of a mechanical system $q = q(t)$ is possible if and only if the variation of the action integral along the curve $q = q(t)$ equals zero."

In other words, the true motion is a critical point of the action function defined on the infinite-dimensional space of smooth mappings of an interval into the configuration space, as shown in Fig. 6.1. Functions on infinite-dimensional spaces are usually called *functionals*. What form does the action functional take for a typical mechanical system?

Consider the motion of a point q on a Riemannian manifold M^n. The trajectory is defined by a mapping $q = q(t)$, $t \in [t_0, t_1]$. We shall assume that the endpoints $q(t_0)$ and $q(t_1)$ are fixed. The tangent vector $\dot{q}(t)$ has the Riemannian square of the length. The quantity $T = (1/2)\dot{q}^2$ is called the *kinetic energy*. We also consider a certain function $U : M^n \to \mathbb{R}$ called the *potential energy*. We set $L = T - U$; this functional is called the *Lagrange function* or the *Lagrangian*.

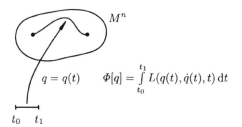

$$q = q(t) \qquad \Phi[q] = \int_{t_0}^{t_1} L(q(t), \dot{q}(t), t)\, dt$$

Fig. 6.1. The true motions are critical points of the action functional

(There was a dispute in the nineteenth century over the notation for co-ordinates and momenta and also over the choice of sign in the Lagrangian. Naturally mathematicians were unhappy with the minus sign. As usual, the physicists triumphed, and the generally accepted notations are now q for co-ordinates, p for momenta, $L = T - U$ for the Lagrangian, $F = \frac{\partial L}{\partial q}$ for force. In any case, physical common sense is satisfied: when a rock is lifted from the surface of the earth, its energy increases. See Klein [2].)

The action function has the form $\Phi[q] = \int_{t_0}^{t_1} L\big(q(t), \dot{q}(t), t\big)\, dt$.

In coordinates the kinetic energy T is a quadratic form: $T = \frac{1}{2} \sum_{ij} a_{ij} \dot{q}_i \dot{q}_j$.
Let us introduce the momenta $p_i = \partial T / \partial \dot{q}_i$.

Geometrically momentum is a linear functional acting on tangent vectors, that is, it is a cotangent vector.

Theorem 1. *The extremals (critical points) of the action functional satisfy the Euler–Lagrange equations*

$$\frac{dp_i}{dt} = \frac{\partial L}{\partial q_i}, \quad i = 1, \ldots, n.$$

(This is a system of n second-order ordinary differential equations for the n unknown functions $q_i(t)$.)

The theorem will be proved below; at this point we shall exhibit the connection the vibrations of a string have with the equation and examine several examples.

Lagrange considered a "discrete string" made up of balls joined together with springs, as in Fig. 6.2.

Fig. 6.2. The Lagrange model of the string

It is not difficult to find the Lagrangian for such a system and derive the equations of motion, and then obtain the equation for the string by passing to the limit, understand its variational basis, and find the Lagrangian. We first consider some simple examples.

Example 1. Suppose a particle is moving freely in a Euclidean space, say the plane for simplicity. Then the potential energy U is zero, and the Lagrangian is $L = T = (1/2)(\dot{q}_1^2 + \dot{q}_2^2)$. It is easy to verify that the extremals of the action are straight lines. Indeed, by Theorem 1 the extremals satisfy the Euler–Lagrange equations, which in the present case have the form $\dot{p}_1 = 0$, $\dot{p}_2 = 0$. But $p_1 = \dot{q}_1$ and $p_2 = \dot{q}_2$, that is, the generalized momenta coincide with the velocities. Then $\ddot{q}_1 = 0$ and $\ddot{q}_2 = 0$. The solutions of these equations are lines, as shown in Fig. 6.3.

Fig. 6.3. Lines are the extremals of the action for a free particle

Example 2. We now introduce a potential energy $U = U(q)$, that is, we let the particle move in a force field. We have the following system of equations:

$$\dot{p}_1 = -\frac{\partial U}{\partial q_1}, \qquad \text{or} \qquad \ddot{q}_1 = -\frac{\partial U}{\partial q_1},$$
$$\dot{p}_2 = -\frac{\partial U}{\partial q_2}, \qquad\qquad \ddot{q}_2 = -\frac{\partial U}{\partial q_2}.$$

Let us consider the case when U is a quadratic form. For example, when $n = 1$ we have $U = aq^2$, and the equation $\ddot{q} = -aq$ is the equation of pendulum motion.

In the general case of a quadratic potential U we refer the quadratic form to principal axes via an orthogonal transformation (in the sense of the metric defined by the kinetic energy). We thereby obtain a system of independent oscillators. Returning to the vibrating string, one can say that it is an infinite system of independent pendulums.

PROOF OF THEOREM 1. For simplicity, to avoid having to write subscripts, we shall assume that $q \in \mathbb{R}$. We have the Lagrangian $L(q, \dot{q}, t)$. Consider a variation of the motion $\delta q(t)$, $\delta q(t_0) = \delta q(t_1) = 0$, as in Fig. 6.4.

Fig. 6.4. The variation of the motion

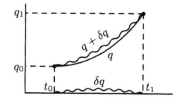

We compute the increment of the Lagrangian:

$$L(q + \delta q, \dot{q} + \delta\dot{q}, t) - L(q, \dot{q}, t) = \frac{\partial L}{\partial q}\delta q + \frac{\partial L}{\partial \dot{q}}(\delta\dot{q}) + o(\delta) .$$

We now find the principal linear part of the increment of the action:

$$\delta \int_{t_0}^{t_1} L(q, \dot{q}, t)\, dt = \int_{t_0}^{t_1} \left(\frac{\partial L}{\partial q}\delta q + \frac{\partial L}{\partial \dot{q}}(\delta\dot{q}) \right) dt \overset{(*)}{=} \int_{t_0}^{t_1} \left(\left(-\frac{d}{dt}\frac{\partial L}{\partial \dot{q}} \right) \delta q + \frac{\partial L}{\partial q}\delta q \right) dt$$

$$= \int_{t_0}^{t_1} \delta q \left(-\frac{d}{dt}\frac{\partial L}{\partial \dot{q}} + \frac{\partial L}{\partial q} \right) dt .$$

The transition marked $(*)$ is an integration by parts, taking account of the fact that the integrated term equals 0, since the variation of the trajectory at the endpoints of the interval is zero.

If the trajectory is critical, this last expression is identically zero. Then

$$-\frac{d}{dt}\frac{\partial L}{\partial \dot{q}} + \frac{\partial L}{\partial q} \equiv 0 .$$

For if not, the variation δq could be chosen so that the integral is not identically zero (this is an exercise for the reader). The theorem is now proved.   □

In the case of free particle motion the Lagrangian has the form $L = T$. It then follows immediately from the Euler–Lagrange equations that the critical trajectories are straight lines. The integral is minimized on them. For the problem of the shortest path, in which $L = \sqrt{2T}$, the extremals are again straight lines, and the minimum is reached on them.

Important remark. This theory is independent of the coordinate system. For that reason, if a certain equation is an Euler–Lagrange equation, it has the form of an Euler–Lagrange equation in any other coordinate system. Hence, it suffices to perform the change of coordinates in the Lagrangian alone.

Exercise. Consider a uniform motion $q(t)$ of particles along lines. Then $q(t)$ is the solution of the equation $\ddot{q} = 0$. Write the corresponding equation in polar coordinates.

For more details on the methods of the calculus of variations, see [1, Ch. 3], which contains further references.

Our theory provides a powerful method of studying systems in a neighborhood of a critical point in relation to the deviation from the equilibrium position of the potential energy, for example, a minimum. We replace the potential by its quadratic part. The error in the right-hand side of the equation (that is, in the force) will be of second order of smallness in comparison with its distance from the equilibrium position. The linear portion does not change when this is done, so that the change just described is simply a linearization. We replace the quadratic form of the kinetic energy, which depends both on a parameter and on the position q, by a constant form taken at the minimum point. In the finite-dimensional case the kinetic and potential energy have the form

$$T = \frac{1}{2}(A\dot{q}, \dot{q}) , \quad U = \frac{1}{2}(Bq, q) .$$

The first form is positive-definite, but the second need not be if the initial critical point of the potential energy is not a minimum.

It is known that two such forms can be simultaneously referred to principal axes. This is done by bringing the second form to principal axes via an orthogonal transformation for which the first form defines the Euclidean structure of the space. To do this in practice it is necessary to solve the characteristic equation $\det(B - \lambda A) = 0$. This equation arose historically in the work of Lagrange on the secular perturbations of the orbits of planets near Keplerian orbits. For that reason the characteristic equation was called the *secular equation*.

Thus in the theory of small oscillations, by definition, the kinetic energy T defines a Euclidean structure in the configuration space and the potential energy U is a quadratic form in that space. In Cartesian coordinates the Euler–Lagrange equations have the form $\ddot{q} = -\nabla U = -Bq$.

There is a simple geometric procedure for finding the principal axes. Consider a level ellipsoid of the quadratic form (Bq, q). The point farthest from the origin in the sense of the given Euclidean structure determines the first eigenvector (principal axis). The other principal axes lie in the orthogonal complement of the first, and this hyperplane intersects the ellipsoid in an ellipsoid of smaller dimension, in which the same procedure can then be applied, and so on, as in Fig. 6.5. It is surprising that this method also works in an infinite-dimensional space.

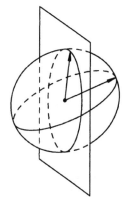

Fig. 6.5. Finding the principal axes

In principal axes the system splits into one-dimensional equations that can be solved explicitly. (We write out the solutions in the case of a positive-definite form B):

$$\ddot{q}_k = -\lambda_k q_k , \quad q_k(t) = a_k \cos \omega_k t + b_k \sin \omega_k t , \quad \text{where } \omega_k = \sqrt{\lambda_k} .$$

These formulas describe the independent natural harmonic oscillations along orthogonal directions. If the frequencies are incommensurable, the motion as a whole is non-periodic.

It is convenient to write the solution corresponding to the natural oscillations in the complex form $q(t) = \text{Re}(A_k e^{i\omega_k t} \xi_k)$. Here A_k is the complex amplitude and ξ_k is the eigenvector. All possible solutions are obtained by adding these solutions: every oscillation is a superposition of natural oscillations. The original system thus turns out to be a system of n noninteracting oscillators.

As the potential energy U increases all the natural oscillations increase (for the proof see [1]). From this we obtain the following simple but amazing facts about the ellipsoids:

- if one ellipsoid lies inside another, each axis of the smaller ellipsoid is less than the corresponding axis of the larger;
- consider a section of the ellipsoid by a plane passing through its center; the axis of the section has a length intermediate between those of the axes of the original ellipsoid, as in Fig. 6.6.

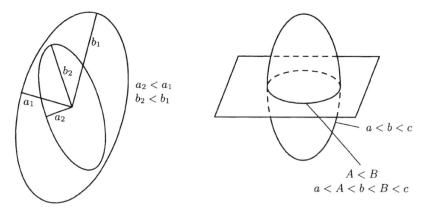

Fig. 6.6. Relations between the axes of ellipsoids

This effect is observed in practice, for example, when a crack in a bell is filled in; the cracked bell has a dull sound, which becomes more resonant when the crack is filled in.

For more details on small oscillations see [1, Ch. 5] and also [3, § 25, par. 6].

To generalize our theory to infinite-dimensional systems we need some groundwork. But first let us see what results are to be obtained.

In the case of a one-dimensional string with endpoints fixed the configuration space is the space of functions $u(x)$: $u(0) = u(l) = 0$. We shall assume that these functions are smooth. The kinetic energy has the form

$$T = \frac{1}{2} \int_0^l \left(\frac{\partial u}{\partial t} \right)^2 dx \ .$$

The potential energy is the work done in bringing the string into the given shape $u(x)$. We shall assume that this form differs only by a small amount from the stationary position $u \equiv 0$. In computing the potential energy in the approximation provided by the theory of small oscillations we need to take account of terms of order u^2, but we can neglect terms of higher order (u^3 and so forth). Since no work is expended in bending the string, all the work goes into lengthening the string from its original length l to a new length, that of the graph of $u(x)$.

Lemma. *The potential energy of a tightly stretched string is proportional to its lengthening (in the approximation of small oscillations) with a coefficient proportional to the tension.*

PROOF. By Hooke's law the tension is proportional to the lengthening. Lengthening is a quantity of second order of smallness relative to the length of the string. Consequently, the tension in the deformed string can be regarded as constant and independent of the shape of the string. (This approximation is valid within the limits of the approximation of small oscillations, that is, when the energy is computed taking account of quantities of second-order smallness relative to the deviation but not third-order. In this approximation it suffices to carry out computations retaining terms of first order of smallness relative to the deviation and neglecting terms of second-order smallness.)

If the tension is constant throughout the deformation, the energy of an element of the string can be computed as the work of this constant force on an element of the path and therefore (in this approximation) proportional to the lengthening of this element (and the magnitude of the tension).

Summing the potential energies of all the elements of the deformed string, we find that in the present approximation its potential energy equals the product of the tension and the lengthening of the entire string.

Neglecting infinitesimals of second order relative to u, we finally obtain the following expression for the potential energy:

$$U = F \int_0^b \left(\sqrt{1 + \left(\frac{\partial u}{\partial x} \right)^2} - 1 \right) dx \approx \frac{F}{2} \int_0^b \left(\frac{\partial u}{\partial x} \right)^2 dx \ . \qquad \square$$

Remark. The same expression could have been derived by considering the model with beads and springs and computing the force acting on the ith bead from sides $(i - 1)$ and $(i + 1)$ (assuming that this bead moves in the direction perpendicular to the unperturbed string). The projections of these two forces in the direction normal to the unperturbed string are in first approximation proportional to the differences in the displacements of successive beads $q_i - q_{i-1}$ and $q_{i+1} - q_i$ respectively (with opposite signs and common coefficient F equal to the tension in the string).

The potential energy $U = \sum \frac{F}{2}(q_{i+1} - q_i)^2$ leads precisely to the following expression for the force acting on the ith bead:

$$-\frac{\partial U}{\partial q_i} = F\big[(q_{i+1} - q_i) - (q_i - q_{i-1})\big] = F(q_{i+1} - 2q_i + q_{i-1}) \,.$$

In the limit the sum becomes the integral obtained in the proof of the lemma.

The Lagrangian L is $T - U$.

To write the Euler–Lagrange equation, let us find the momenta and forces.

Obviously $p(x) = \partial u/\partial t$ (assuming the string is homogeneous and has density 1 throughout).

The potential energy in the pre-limit case equals $\frac{F}{2}\sum(q_{i+1} - q_i)^2$, where $q_i = u(x_i)$ as in Fig. 6.7.

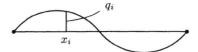

Fig. 6.7. The discrete model of the string

To find the force we take the derivative with respect to q_i, obtaining

$$-F(q_{i+1} - q_i) + F(q_i - q_{i-1}) = -F(q_{i+1} - 2q_i + q_{i-1}) \,.$$

The expression in parentheses on the right is precisely the second difference (Fig. 6.8), which becomes the second derivative $\partial^2 u/\partial x^2$ upon passage to the limit.

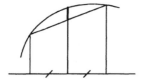

Fig. 6.8. The second difference

Thus, after a second passage to the limit the Euler–Lagrange equation will have the form

$$\frac{\partial^2 u}{\partial t^2} = k\frac{\partial^2 u}{\partial x^2} \,,$$

that is, the form of the wave equation. The coefficient k is directly proportional to its tension and inversely proportional to its density. We already have a method of solving it – finding its natural oscillations. It is necessary to find the principal axes of the quadratic form of the potential energy in the metric given by the kinetic energy. In the finite-dimensional case we wrote:

$$q(t) = Ae^{i\omega t}\xi .$$

Thus for the string it is necessary to seek a solution of the form $u = e^{i\omega t}\xi(x)$.

We substitute this expression into the equation:

$$-\omega^2 e^{i\omega t}\xi(x) = ke^{i\omega t}\xi''(x) ,$$

thus arriving at the eigenfunction problem $\xi''(x) = -\lambda\xi$, $\lambda = \omega^2/k$.

This application of the general theory of oscillations is called the *method of separation of variables*. The solution is:

$$\xi = a\cos\sqrt{\lambda}x + b\sin\sqrt{\lambda}x ; \quad \xi(0) = 0 \Rightarrow a = 0 ;$$
$$\xi(l) = 0 , \quad \sin\sqrt{\lambda}l = 0 , \quad \sqrt{\lambda}l = n\pi ; \quad \sqrt{\lambda} = \frac{n\pi}{l} ; \quad \omega \sim n .$$

We see that there is a countably infinite set of principal axes and eigenfunctions; as the number n increases the natural frequencies increase also. When an eigenfunction is multiplied by the real part of a complex exponential function of time we obtain a standing wave (Fig. 6.9): the shape of the wave does not change, but oscillations harmonic in t occur.

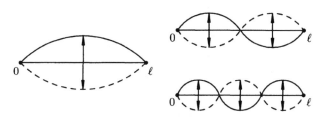

Fig. 6.9. Standing waves

In the finite-dimensional case any oscillation could be decomposed into superpositions of natural oscillations. Such is the case also in the equation of the vibrating string, but it requires a separate proof. (This proof is provided by the theory of Fourier series, which was created for this purpose. For that reason the theory of Fourier series and its generalizations are also called *harmonic analysis*.)

A justification for the possibility of applying the theory of oscillations to systems with an infinite number of degrees of freedom (like the string) was obtained by mathematicians only rather recently (in the late nineteenth and early twentieth centuries), and it led to the creation of functional analysis and then to quantum mechanics. However, the most general algebraic structure of the theory of oscillations with both a finite and an infinite number of degrees of freedom is a much more fundamental theory (used by physicists with complete success long before its rigorous mathematical justification).

The situation here resembles the situation in regard to justifying the theory of real numbers (obtained, strictly speaking, only in the nineteenth century). The ancient Greeks of Pythagoras' time were compelled to conceal the theorem on the incommensurability of the diagonal and side of a square, which had destroyed their belief in the power of numbers (which at the time were understood only as rational numbers).

This did not prevent Newton from creating analysis without bothering about the details of justifying the arithmetic of real numbers (which, however, he knew very well). The author of the present course has attempted to teach the audience how to conjecture and predict bold generalizations (such as in the passage from oscillations with a finite number of degrees of freedom to the oscillations of continuous media) rather than the difficult skill of giving a rigorous justification of these results (including the unavoidable investigation of the minimum necessary smoothness of the objects in question).

At present quantum field theory finds itself in a position analogous to that of the theory of oscillations of continuous media and the theory of real numbers. It provides very impressive mathematical results, but not their justification.

Problem*[1] (Sturm, Hurwitz). Consider a real 2π-periodic function

$$f(x) = \sum_{k \geq N} (a_k \cos kx + b_k \sin kx) \, ,$$

whose Fourier series begins with the Nth harmonics. Prove that the number of zeros of f on the circle $\{x \bmod 2\pi\}$ is not less than the number of zeros of the first harmonic with nonzero coefficient in the series. (That is, it is at least $2N$.)

Example. When $N = 1$, the number of zeros is at least two. This is Morse's inequality: a function on the circle has at least two critical points.

Literature

1. Arnold, V.I.: Mathematical Methods of Classical Mechanics, 2nd edition. Springer, New York (1989)
2. Klein, F.: Vorlesungen über die Entwicklung der Mathematik im 19. Jahrhundert. Chelsea, New York (1964)
3. Arnold, V.I.: Ordinary Differential Equations. Springer, Berlin (1992)

---

[1] From now on, the superscript * in similar situations denotes more tough problems whose solutions are not supplied therein.

Lecture 7

The Theory of Oscillations. The Variational Principle (Continued)

Thus the kinetic and potential energies of the string (Fig. 7.1) have the form

$$T = \frac{1}{2} \int_0^l \dot{u}^2 \, \mathrm{d}x, \quad U = \frac{k}{2} \int_0^l (u_x)^2 \, \mathrm{d}x.$$

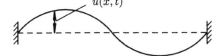

Fig. 7.1. The vibrating string

The Lagrangian L equals $T - U$. The variational principle $\delta \int L \, \mathrm{d}t = 0$ implies the Euler–Lagrange equation $\dfrac{\mathrm{d}}{\mathrm{d}t} \dfrac{\partial L}{\partial \dot{q}} = \dfrac{\partial L}{\partial q}$, which in our case has the form of the vibrating string equation:

$$u_{tt} = k u_{xx} \, .$$

These results can be generalized to the multidimensional case. Let the vibrating body be of arbitrary dimension. For example, when $n = 2$ it is a membrane. The membrane, like the string, can be pictured as the limit of a system of vibrating balls connected with springs, as in Fig. 7.2.

For a membrane or a multidimensional body the kinetic energy can be expressed in analogy with that of the string: $T = \frac{1}{2} \iint (\partial u / \partial t)^2 \, \mathrm{d}x$.

By the integral here we mean the corresponding multidimensional integral.

We remark that one can consider vibrations not only in one direction, but also in more than one. In that case u becomes a vector-valued function. (Geometrically the equilibrium position corresponds to the zeroth section of the bundle, but the bundle itself can be arbitrary.) The kinetic energy will be written by the same formula, except that the integral will be the square of the norm of the velocity vector.

string

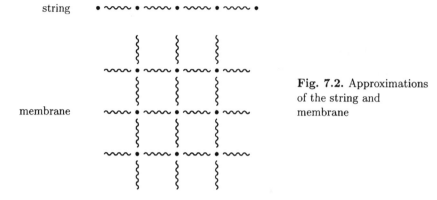

Fig. 7.2. Approximations
of the string and
membrane

membrane

How is the potential energy expressed? Here is the answer:

$$U = \frac{k}{2} \int (\nabla u)^2 \, dx \; ,$$

where $\nabla u = \frac{\partial u}{\partial x} = \left(\frac{\partial u}{\partial x_1}, \ldots, \frac{\partial u}{\partial x_n} \right)$ is the gradient. This integral is called the *Dirichlet integral.*

Dirichlet studied mostly number theory, but he is the one who noticed that harmonic functions minimize this integral (under suitable boundary conditions). This physical idea – the so-called *Dirichlet principle* – has turned out to be a very powerful tool both for proving the existence of solutions of the corresponding problems, but also for studying them and even for practical approximate computation. It is interesting that attempts to prove the Riemann conjecture on the zeros of the zeta function are connected with this integral and its generalizations. The conjecture is that all the non-trivial complex zeros of the zeta function lie on a single real line. The idea (which goes back at least to Hilbert) is to find an oscillation problem through whose eigenfunctions the zeros of the zeta function can be expressed. The fact that the eigenvalues are real would then imply that all the zeros lie on a single real line in the complex plane.

Problem. Find the minimum of the Dirichlet integral in the space of smooth functions on the sphere.

SOLUTION. For a constant the integral equals zero, while for a nonconstant function it is positive. Consequently, the minimum equals zero and is attained only on constant functions.

In the multidimensional case the Euler–Lagrange equations have the form $\frac{\partial^2 u}{\partial t^2} = k \Delta u$, where $\Delta = \frac{\partial^2}{\partial x_1^2} + \cdots + \frac{\partial^2}{\partial x_n^2}$ is the Laplacian in Cartesian coordinates. Actually the Laplacian depends not on the coordinates but on the Euclidean structure of the space, as the next result shows.

Theorem 1. $\Delta u = \operatorname{div}(\operatorname{grad} u)$.

Before proving the theorem we recall the definitions of the concepts that occur here and verify that they are invariant.

Let u be a function. Its differential du is a 1-form acting on a tangent vector ξ: $du|_x(\xi) = (a_x, \xi)$, since any 1-form in Euclidean space is the inner product with a fixed vector at the given point. This vector is called the *gradient* of the function u at the point x: $a_x = \operatorname{grad} u|_x$. Thus the gradient is determined by the Euclidean structure of the space (or more generally, by the Riemannian metric of a manifold).

In any coordinates we have $du = \left(\frac{\partial u}{\partial x_1} dx_1 + \cdots + \frac{\partial u}{\partial x_n} dx_n\right)(\xi) = \sum \frac{\partial u}{\partial x_i} \xi_i$. In orthonormal Cartesian coordinates this is the notation for the inner product of the vectors ξ and $\left(\frac{\partial u}{\partial x_1}, \ldots, \frac{\partial u}{\partial x_n}\right)$, so that in orthonormal Cartesian coordinates the gradient of u has the components $\left(\frac{\partial u}{\partial x_1}, \ldots, \frac{\partial u}{\partial x_n}\right)$.

Exercise. Find the components of $\operatorname{grad} u$ in polar coordinates on the Euclidean plane and in spherical coordinates in three-dimensional Euclidean space.

The divergence of a vector field is defined on a manifold on which a volume element is defined, in particular, on a Riemannian manifold.

Consider a vector field $v = \sum v_i(x)\frac{\partial}{\partial x_i}$ and the corresponding equation $\dot{x} = v(x)$ of the phase flow in n-dimensional space.

Let us recall the definition of the flux of a field across a surface. We denote the volume element by $\tau = \tau^n$. The volume element τ^n is an n-form. Consider the $(n-1)$-form $i_v\tau$ corresponding to the vector v. This form is obtained by substituting the vector v in the first argument of τ^n and leaving the other $n-1$ arguments free.

The *flux* of the field v across an $(n-1)$-dimensional oriented hypersurface is the integral of this form over the surface.

Example. Suppose $\tau = dx_1 \wedge \cdots \wedge dx_n$ in orthonormal Cartesian coordinates. Then

$$i_v\tau = v_1\, dx_2 \wedge \cdots \wedge dx_n - v_2\, dx_1 \wedge dx_3 \wedge \cdots \wedge dx_n + \cdots \pm v_n\, dx_1 \wedge \cdots \wedge dx_{n-1}.$$

In particular, when $n = 3$ for the velocity field of the system $\dot{x} = P, \dot{y} = Q, \dot{z} = R$, we obtain $i_v\tau = P\, dy \wedge dz + Q\, dz \wedge dx + R\, dx \wedge dy$.

The flux has a clear meaning in fluid dynamics. It represents the amount of the phase fluid passing through the surface in unit time, as in Fig. 7.3.

Fig. 7.3. The flux of a vector field through a surface

Consider the flux $\bar{V}_\varepsilon$ of the field v through a small sphere of radius ε with center at x. The *divergence* of the field v at the point x is the limit $\lim\limits_{\varepsilon \to 0} \bar{V}_\varepsilon / \tau(\varepsilon)$, where $\tau(\varepsilon)$ is the volume of the ball. The meaning of the divergence is the "density of the source" at the given point.

Thus the divergence is defined if a volume element is given. In particular the divergence is defined in Riemannian and Euclidean spaces. The divergence of a field v on a manifold with volume element τ is connected with the exterior differentiation of $(n-1)$-forms by the relation

$$(\operatorname{div} v)\tau = \mathrm{d}(i_v \tau) ,$$

which is completely independent of the coordinate system.

Theorem 2. *In orthonormal Cartesian coordinates the divergence of the field v is the trace of the Jacobian matrix of v:*

$$\operatorname{div} v = \frac{\partial v_1}{\partial x_1} + \cdots + \frac{\partial v_n}{\partial x_n} .$$

PROOF. Consider the new position of the point x under motion along the phase flow of the field v over a small time interval of length ε: $x \mapsto x + \varepsilon v(x) + o(\varepsilon)$. Thus the volume element undergoes a small linear transformation, as in Fig. 7.4.

Fig. 7.4. A small linear transformation of the volume element

Lemma. *In first approximation only the changes in vectors in their own direction contribute to the change in the volume of the parallelepiped they span.*

Exercise. Prove the lemma.

Since the parallelepiped spanned by n vectors can be taken instead of a sphere in the definition of the divergence, when computing the flux one must find the increment of the volume of this parallelepiped under a small translation along v. According to the lemma the principal term of this increment is the trace of the Jacobian multiplied by the volume of the original parallelepiped. Dividing by that volume and passing to the limit, we find that the divergence equals the trace. Theorem 2 is now proved. □

In Cartesian coordinates we find by an elementary computation that $\Delta u = \operatorname{div}(\operatorname{grad} u)$. The Laplacian is defined by this formula in any Riemannian manifold.

Problem. Compute the Laplacian of a function defined on the circle in the Euclidean plane.

ANSWER. $\Delta u = \dfrac{\partial^2 u}{\partial \varphi^2}$, where $x = \cos \varphi$ and $y = \sin \varphi$.

The formula $\Delta u = \dfrac{\partial^2 u}{\partial x_1^2} + \cdots + \dfrac{\partial^2 u}{\partial x_n^2}$ is valid only in orthonormal Cartesian coordinates. The operator has different expressions in other coordinates, even when the space is Euclidean.

Exercise. Write the Laplacian in polar, cylindrical, and spherical coordinates.

Remark. As will be shown below, the geometric meaning of the Dirichlet integral is the principal term of the increment in area of the membrane. It is remarkable that it is a quadratic expression.

This is completely obvious in the one-dimensional case: when the angle is small, the difference between the hypotenuse and a leg is of second-order smallness (verify this), as shown in Fig. 7.5. This simple fact has great consequences and important applications.

$$b = a\varepsilon + O(\varepsilon^2), \quad c = a + O(\varepsilon^2)$$

Fig. 7.5. The hypotenuse is practically equal to the leg

Examples. 1. If you return home along a sinusoidal curve you increase the distance you must walk only slightly (by approximately 20%), since only the portions of the curve with steep slope make a significant contribution to the increase in distance, and these are only a small portion of the curve, as shown in Fig. 7.6.

Fig. 7.6. A sinusoidal curve is not much longer than a straight line

2. By inclining the motor slightly from the longitudinal axis of an airplane one can prevent the tail assembly from burning up in the wake, with a loss of thrust of second-order smallness, as shown in Fig. 7.7. For example, even at a large inclination of 6°, we obtain $\varepsilon \approx 0.1$, which yields a loss in thrust of only 0.5%.

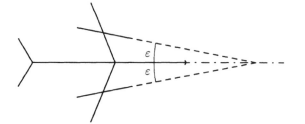

Fig. 7.7. Rotating the motor makes practically no change in the thrust

3. By analyzing the results of Tycho Brahe's astronomical observations[1] Kepler at first believed that Mars moved around the Sun in a circle, but that the Sun was not at the center. Indeed, if we consider an ellipse with semiaxes a and b and small eccentricity e, as in Fig. 7.8, we have $b = a\sqrt{1 - e^2} = a(1 - e^2/2 + \cdots)$. When e is small (about $1/10$ for Mars), it is difficult to observe the difference between the ellipse and a circle, even though the difference between the center and the focus is noticeable.

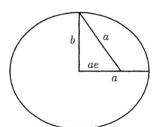

Fig. 7.8. An ellipse with small eccentricity is almost indistinguishable from a circle

This can also be confirmed by the following simple experiment. Add one drop to a circular cup of tea near the center. After reflecting from the walls of the teacup, the waves will converge to a point symmetric with respect to the center, as in Fig. 7.9.

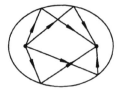

Fig. 7.9. Convergence of waves in a disk and in an ellipse

---

[1] These observations were made with his unaided eye. Even at the end of the seventeenth century, it was still remained to be proved that observations made through a telescope gave precision at least as good.

EXPLANATION. If a drop is added precisely at the focus in an elliptical teacup, the waves will converge at the other focus after reflection. A circle differs by only a small amount from an ellipse with small eccentricity. For that reason the waves converge at the second focus of this approximate ellipse (not precisely, but slightly spread out). With careful observation one can detect a second convergence of the waves at the original point.

We can now explain why the difference is of second order of smallness. A line minimizes the length functional, so that the increment cannot be of first-order smallness.

Let us now return to the geometric meaning of the Dirichlet integral. It is asserted that this integral is the principal part of the increment of area, that is, if S_0 is the area of the membrane in equilibrium and S_ε its area at the displacement εu, we need to prove that the new area S_ε equals $S_0 + \frac{\varepsilon^2}{2} \int (u_x)^2 \, dx + o(\varepsilon^2)$, as illustrated in Fig. 7.10. Let us look at the change in area of a small region. We choose a convenient coordinate system: one axis goes along $\operatorname{grad} u$ and the other is orthogonal to it, as shown in Fig. 7.11. Then the new area is the product of the new lengths, one of which has not changed, while the other is computed as in the one-dimensional case. As a result we have $S_\varepsilon = S_0(1 + (\varepsilon^2/2)(\nabla u)^2 + \cdots)$.

Fig. 7.10. Change in the area of a membrane

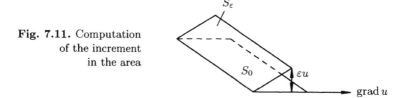

Fig. 7.11. Computation of the increment in the area

It remains to integrate this increment over the entire membrane.

We can carry out all computations in the pre-limiting system of balls and springs shown in Fig. 7.12.

We construct a homotopy αu from the equilibrium position to the displacement εu; in doing so we need to find the term of order ε^2 in the expression for the potential energy, that is, the work done in making such a

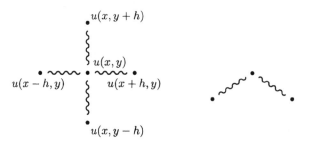

Fig. 7.12. Computation of the potential energy of a membrane

displacement. The side view in both orthogonal directions is the same as in the one-dimensional case (Fig. 7.12). Hence the balancing forces are already computed and consist of the second differences with a negative sign (with a suitable choice of units of measurement):

$$-u(x - h, y) + 2u(x, y) - u(x + h, y) , \quad -u(x, y - h) + 2u(x, y) - u(x, y + h) .$$

By passing to the limit we obtain terms of the form $-u_{xx}$ and $-u_{yy}$. Integrating with respect to α from 0 to ε, we obtain

$$-\int \alpha \Delta u \, d\alpha u = -u \Delta u \int \alpha \, d\alpha = -\frac{\varepsilon^2}{2} u \Delta u .$$

Finally, by integrating over the region we obtain (up to a numerical multiple)

$$U = -\iint \left(\frac{\partial^2 u}{\partial x^2} + \frac{\partial^2 u}{\partial y^2} \right) u \, dx \, dy .$$

We now integrate by parts (first with respect to x)

$$\int \frac{\partial^2 u}{\partial x^2} u \, dx = -\int \frac{\partial u}{\partial x} \frac{\partial u}{\partial x} \, dx + u \frac{\partial u}{\partial x} \Big|_{x_1}^{x_2} .$$

We now take into account the fact that $u = 0$ on the boundary (the membrane is clamped). Thus the integrated term vanishes. The integration with respect to y is similar. As a result, we find

$$U = \iint (\nabla u)^2 \, dx \, dy .$$

Thus the potential energy is proportional to the increment of area. As we have seen, the Euler–Lagrange equation for this case is the wave equation.

In the course of the proof we have proved the formula

$$\int (\nabla u)^2 \, dx = -\int \Delta u \, u \, dx \quad \left(u \big|_{\partial \Omega} = 0 \right) .$$

In addition, we have computed the force, which turned out to be $-\Delta u$, so that the wave equation can be interpreted as a Newtonian equation. But we can legitimately compute $\frac{\partial L}{\partial q}$ once we know T and U. The latter are quadratic forms on an infinite-dimensional function space ($\dot{q} = u_t$, $q = u$).

Let us find the variation $\delta \int \frac{1}{2}\left(\frac{\partial u}{\partial x}\right)^2 \, \mathrm{d}x$. Let $u = u_0(x) + \varepsilon\xi(x)$, $\xi\big|_{\partial\Omega} = 0$.

We need to find the coefficient of ε in the expansion of this integral and represent it as an infinite-dimensional inner product:

$$\int \frac{1}{2}\left(\frac{\partial u_0}{\partial x} + \varepsilon\frac{\partial \xi}{\partial x}\right)^2 \mathrm{d}x = \int \frac{1}{2}\left(\frac{\partial u_0}{\partial x}\right)^2 \mathrm{d}x + \varepsilon \int \frac{\partial u_0}{\partial x}\frac{\partial \xi}{\partial x}\,\mathrm{d}x + \cdots$$

$$= -\varepsilon \int \xi\,\Delta u \, \mathrm{d}x + \int \frac{1}{2}\left(\frac{\partial u_0}{\partial x}\right)^2 \mathrm{d}x + \cdots$$

(in the integration by parts we used the fact that $\xi\big|_{\partial\Omega} = 0$). The first of these integrals happens to be the inner product. Hence the force is indeed $-\Delta u$.

All this reasoning is also valid for a Riemannian manifold.

It is a remarkable fact that a large class of problems is described by the wave equation. The wave equation can be derived starting from the assumption that the given variational problem satisfies certain conditions of an axiomatic nature. (To be sure, the variational derivation of the problems of mathematical physics remains without any convincing explanation.) The axioms are the following:

1. The kinetic energy T is a quadratic form in the velocities, which entails the *spatial homogeneity* of the system. However, when the density is variable, T is still expressed by a similar form.

We confine ourselves to the case of a homogeneous medium, in which T is the integral of $\frac{1}{2}\left(\frac{\partial u}{\partial t}\right)^2$. We shall exhibit the axioms that U must satisfy in order for the problem to be described by a wave equation.

2. The quadratic form U is additive and local, which physically means there is no distant action. The quadratic form can be written as $U = \frac{1}{2}(Au, u)$, where A is a linear operator and the parentheses denote the inner product (that is, the integral of the product of the functions). An operator A is *local* if the value of Au at a point can be computed in terms of a finite number of derivatives of the function at that point.

In our case only neighboring points interact, which leads to the appearance of only first-order derivatives u_x in the Dirichlet integral (and only second derivatives in the corresponding symmetric operator $A = -\Delta$). The interaction of a large number of points leads to the appearance of higher-order derivatives. For example, the operator $\Delta^2 u$ appears in the equation for the problem of the bending of a thin plate. In the case of a membrane the increase in energy occurs only because of the stretching, not bending. For a plate energy is expended in bending as well. This is also an example of a local system, but with higher-order derivatives. If only first-order derivatives occur in the expression for the potential energy, it will have the general form

$$\sum a_{ij}\frac{\partial u}{\partial x_i}\frac{\partial u}{\partial x_j} + \sum b_j\frac{\partial u}{\partial x_j}u + cu^2 .$$

In which cases does the corresponding symmetric operator become the Laplacian? When further restrictions are imposed on the operator acting on the functions in the Euclidean space $\mathbb{R}^n$:

3. Homogeneity (translation-invariance):

$$\Delta\big(u(x+a)\big) = (\Delta u)(x+a) .$$

4. Isotropy (rotation-invariance):

$$\Delta\big(u(gx)\big) = (\Delta u)(gx) .$$

The quadratic form in the gradient vector defines a level ellipsoid in the tangent space at each point. The space is isotropic if this ellipsoid is a sphere, that is, the properties of the membrane are independent of rotation. Non-isotropic media are also encountered.

It follows from the homogeneity and isotropy that $b = 0$, since the linear part of the operator is the inner product of b and $u\frac{\partial u}{\partial x}$. If b were nonzero, the properties of the membrane would change under some rotation.

The term cu^2 may occur in a homogeneous isotropic medium. For example, it occurs in the problem of the oscillations of a balloon (thin shell) about an equilibrium position. The absence of this term is caused by the following restriction.

5. The metric of the space (x, u) is invariant under translations along the u-axis.

All these restrictions lead to a situation in which only the Laplacian remains in the equation.

Now let us consider the problem of the equilibrium positions – the critical points of the potential energy. If $u\big|_{\partial\Omega} = 0$, then 0 is a stable equilibrium position, a minimum of the Dirichlet integral. But other boundary conditions can be considered. For example, consider a string with endpoints clamped at arbitrary points, as in Fig. 7.13.

Fig. 7.13. The Dirichlet problem for the vibrating string

In this case the minimum of the potential energy is attained on a line.

We now pose the analogous problem for the vibrating membrane: $\Delta u = 0$, $u\big|_{\partial\Omega} = \varphi$. This is the *Dirichlet problem for Laplace's equation*, shown in Fig. 7.14. The solutions of the equation $\Delta u = 0$ are called *harmonic functions in the domain* Ω. We shall seek a solution of the Dirichlet problem in the class of functions that are continuous on the closure of Ω and twice differentiable in its interior.

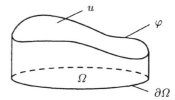

Fig. 7.14. The Dirichlet problem for the vibrating membrane

There is a general device for finding a minimum of the potential U. To do this one must move along the gradient, that is, along the vector field $\dot q = -\nabla U$. When the quadratic form U is positive-definite, we necessarily reach a minimum in this way. In our case the equation for gradient descent has the form $(\dot q =) \frac{\partial u}{\partial t} = k\,\Delta u \,(= -\operatorname{grad} U)$. In particular, for the string we obtain the equation $\frac{\partial u}{\partial t} = k u_{xx}$, which is called the *diffusion equation* or the *heat equation*.

The heat equation is obtained in the problem of heat propagation. Consider elements being heated that are located at the nodes of a lattice. At each successive moment of time the value of the temperature at a node is some weighted mean of its values at five points, as in Fig. 7.15. For simplicity we draw the one-dimensional case (also Fig. 7.15).

Taking the average, we obtain a new function, to which the same procedure is applied, and so forth. This is precisely an implementation of gradient descent. Thus the propagation of heat occurs just as a minimum of potential energy can be sought using the gradient method.

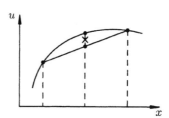

Fig. 7.15. Equalization of the temperatures at the nodes

If the equation is interpreted as a diffusion equation, u corresponds to the density of the distribution of particles. The Laplacian arises as the result of the invariance conditions for the system described above. Incidentally, in this case the term cu in the equation (corresponding to $cu^2/2$ in the potential energy) is often physically justified: it corresponds to the creation or annihilation of particles, depending on the sign of c.

A method of finding stationary solutions (equilibrium positions) has thus been described.

In special cases it is possible to find explicit solutions. We first study harmonic functions. Being solutions of the linear equation $\Delta u = 0$, they form a vector space. Let us attempt to satisfy the boundary conditions. When $n = 1$ harmonic functions are linear functions and the space of them is two-dimensional. A basis is formed by the functions 1 and x. One can pass a (unique) line through two given points and satisfy the boundary conditions.

Now let $n = 2$.

Theorem 3. *Let us identify the real plane with the complex line. Let $f(z)$ be a holomorphic function. Then $\operatorname{Re} f$ and $\operatorname{Im} f$ are harmonic functions.*

Exercise. Prove the theorem. Use the Cauchy–Riemann equations.

Problems. 1. Let $\varphi(x, y)$ be the oriented angle subtended by a fixed line segment in the plane when seen from the point (x, y). Prove that $\varphi(x, y)$ is a harmonic (multi-valued) function, except at the endpoints of the line segment.

2. Construct a bounded function that is harmonic inside a disk, continuous everywhere except at two given points on the boundary, equal to 0 on one of the two arcs into which the two points divide the boundary, and equal to 1 on the other arc.

3. Carry out the construction of the preceding problem for the case when the circle is divided into n arcs and n boundary values are given.

4. Find all the harmonic functions on the sphere, that is, solve the equation $\operatorname{div} \operatorname{grad} u = 0$, where u is a smooth function on the sphere in $\mathbb{R}^n$.

HINT. If a function is not a constant, its Dirichlet integral is nonzero, and the function cannot be a critical point of the Dirichlet integral. Therefore $\Delta u \neq 0$.

5. A function is given at time 0 on the vertices of a cube. At each successive time $(i + 1)$ the value at each point is replaced by the arithmetic mean of the values at the adjacent vertices at the preceding time (i). Find the limit of the resulting sequence of functions as $i \to +\infty$.

HINT. The limiting function is "harmonic."

6. Let u be a homogeneous function of degree 0 on $\mathbb{R}^n$, that is, $u(\lambda x) = u(x)$ for any x and any $\lambda > 0$. Prove that

$$r^2 \Delta u = \tilde{\Delta} u \,,$$

where $\tilde{\Delta}$ is the Laplacian $\operatorname{div} \operatorname{grad}$ on the unit sphere, that is, $(\tilde{\Delta} u)_{r=1} = \operatorname{div} \operatorname{grad}\left(u\big|_{r=1}\right)$.

7. Let u be a homogeneous function of degree k in $\mathbb{R}^n$, that is $u(\lambda x) = \lambda^k u(x)$ for any x and $\lambda > 0$. Prove that

$$\tilde{\Delta} u = r^2 \, \Delta u - (k^2 + (n-2)k)u \ ,$$

where $\tilde{\Delta} u$ is a homogeneous function of degree k coinciding with the function $\operatorname{div} \operatorname{grad}(u|_{r=1})$ when $r = 1$.

In particular, for homogeneous functions of degree k on the plane $(n = 2)$

$$\tilde{\Delta} u = r^2 \, \Delta u - k^2 u \ ,$$

and for homogeneous functions of degree k in three-dimensional space

$$\tilde{\Delta} u = r^2 \, \Delta u - (k^2 + k)u \ .$$

8. Find all homogeneous harmonic functions in $\mathbb{R}^n \setminus \{0\}$ depending only on r.

HINT. The case $n = 2$ is special.

9*. Prove that the number of domains into which the zero level curve of the kth eigenfunction of the Laplacian in a complex domain (with Dirichlet boundary conditions) divides this domain is at most k. This result holds on compact Riemannian manifolds of any dimension, for example, on spheres.

For example, the first eigenfunction (having eigenvalue of smallest absolute value) has no changes of sign at all (it vanishes only on the boundary).

HINT. The ratio of the Dirichlet integral to the integral of the square of an eigenfunction is the same on each of the N components of the complement of the zero level curve of the eigenfunction. Hence there exists an N-dimensional space of functions on which this ratio of quadratic forms is the same as for the eigenfunction.

It follows from the theorems of Lecture 6 on the axes of an ellipsoid that the index of axis k is not less than the dimension N.

10. Consider the eigenfunctions of the Laplacian on the n-dimensional torus:

$$\sum \frac{\partial^2 u}{\partial x_j^2} = -\lambda u \ , \quad (x_j \bmod a_j) \in T^n \ .$$

We denote by $N(E)$ the number of eigenfunctions for which $\lambda \leq E$. Study the behavior of $N(E)$ as $E \to \infty$.

HINT. The Laplacian commutes with translations. Therefore the eigenfunctions are exponentials

$$e_k(x) = e^{2\pi i (k,x)} \ , \quad (k,x) = \sum k_j x_j / a_j \ , \quad k_j \in \mathbb{Z} \ .$$

Consequently the question reduces to computing the number of points with integer coordinates in a large ellipsoid.

ANSWER. In the cotangent bundle to the torus T^*T^n consider the domain $\Omega(E)$ defined by the inequality $\sum \xi_j^2 \leq E$ (ξ_j are the coordinates of the form $\sum \xi_j \, dx_j$). Then as $E \to \infty$

$$N(E) \sim (2\pi)^{-n} \operatorname{Vol} \Omega(E) = \text{const} \cdot E^{n/2} \,.$$

Remark 1. The analogous *Weyl formula* holds on any manifold and for any "elliptic" operator, for example, for an operator of second or higher order with variable coefficients $P\left(x, i\frac{\partial}{\partial x}\right) = \lambda u$. Here the domain Ω is defined by the condition $\widetilde{P}(x, \xi) \leq E$, where $\widetilde{P}$ is the sum of the terms of the polynomial P that are of highest order with respect to ξ (the "principal symbol").

Remark 2. The later terms of the asymptotic expansion of $N(E)$ are difficult to study, even for the usual Laplacian when $a_j = 2\pi$. The problem is that the surface of the sphere $\sum k_j^2 = E$ may contain many points with integer coordinates (in other words, the eigenvalue E may be of large multiplicity). This possibility occurs, for example, in the case of the Laplacian on the sphere S^2, for which the eigenvalue E has multiplicity of order $\sqrt{E}$ (see Lecture 11).

Lecture 8

Properties of Harmonic Functions

In this lecture we shall study the properties of harmonic functions, mostly in the plane, although some theorems will be proved in the n-dimensional case. We shall begin with the three-dimensional case.

Consider the function $u = 1/r$, where $r = \sqrt{x^2 + y^2 + z^2}$ and (x, y, z) are Cartesian coordinates.

Exercise. Verify that the function u is harmonic.

This simple example is actually very important. The reason is that the two most important forces studied in physics – gravitational and electrostatic – are structured as follows: they act along the line joining the particles and their magnitude is inversely proportional to the square of the distance between the particles.

But in that case the potential whose gradient is the force of attraction to one particle is proportional to $1/r$, which is a harmonic function. A linear combination of harmonic functions is harmonic, so that the potential of the force of attraction toward a finite number of particles is also harmonic.

From a finite number of particles one can pass to a continuous mass (or charge) distribution with density $\rho(x)$ in a region D. Then the potential of the field of attraction toward the body in question at a point x is computed by the formula

$$\int_D \frac{\rho(y)\, dy}{|x - y|} \, ,$$

as shown in Fig. 8.1, the function being a harmonic as a function of x in a region free of mass (resp. charge). Thus the fundamental laws of nature are connected with harmonic potentials.

Let us attempt to understand why the function $u = 1/r$ is harmonic, that is, why $\operatorname{div} \operatorname{grad} u = 0$, without doing any computation.

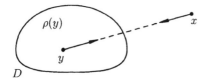

Fig. 8.1. A gravitational field

First of all, at the point x, div grad u depends only on r, the distance from x to 0. Thus div grad $u(x) = f(\|x\|)$. Now consider the region between concentric spheres of radius r and R, with $r < R$. Let us compute the flux of grad u across the boundary of this region. The gradient of u is directed toward 0 and hence is perpendicular to these spheres. The area of the outer sphere is $(R/r)^2$ times as large as that of the inner sphere. But the magnitude of the gradient on the outer sphere is less than on the inner sphere by exactly this same factor (since the force is inversely proportional to the square of the distance). Therefore the fluxes of the gradient across the two spheres are exactly equal, so that the total flux of the gradient across the boundary of the spherical shell is exactly 0. It follows that div grad$(1/r)$ vanishes. Of course, the equality $\Delta(1/r) = 0$ can also be verified by simply differentiating.

We now turn to the case of the plane. Consider a holomorphic function $f(z)$ on the plane or in a domain of the plane. Then from $f(z)$ one can easily form the harmonic functions Re $f(z)$, Im $f(z)$, $\ln|f(z)|$, and arg $f(z)$ (the last two being the real and imaginary parts of the holomorphic function $\ln f(z)$), and the like. The converse is also true. Every harmonic function can be obtained in this way.

Theorem 1. *Every harmonic function in a simply connected domain of the plane is the real part of a holomorphic function, which is defined in that domain up to an additive purely imaginary constant.*

PROOF. We are seeking a holomorphic function $f = u + iv$, the harmonic function u being given and v to be found. By the Cauchy–Riemann equations $u_x = v_y$ and $u_y = -v_x$, so that $\mathrm{d}v = \alpha$, where $\alpha = (-u_y)\,\mathrm{d}x + (u_x)\,\mathrm{d}y$. In a simply connected domain there exists a function v with this differential provided $\mathrm{d}\alpha = 0$. In the present case we can define v to be $\int_{x_0}^{x} \alpha$, and the integral will be independent of the path within the domain, since by Green's theorem the integral over any closed path is zero. Thus it remains only to verify that $\mathrm{d}\alpha = 0$. But $\mathrm{d}\alpha = (u_{xx} + u_{yy})\,\mathrm{d}x \wedge \mathrm{d}y = 0$ since u is harmonic. Thus the function v exists and is determined up to a real additive constant. □

We see that the theories of harmonic and analytic functions in the plane essentially coincide, the former being the real parts of the latter. This fact determines the role played by analytic functions in mathematical physics. By use of these functions one can obtain the solutions of many problems, for example, airflow around a wing (a Zhukovskiĭ airfoil).

Theorem 2 (the mean-value theorem). *The mean value of a harmonic function over a circle equals its value at the center (see Fig. 8.2):*

$$\frac{1}{2\pi} \int_0^{2\pi} u \, d\varphi = u(0) \, .$$

Fig. 8.2. The mean-value theorem

We shall prove the analogous theorem in the multidimensional case below, and we shall see that this property can be taken as the definition of a harmonic function.

PROOF. As we know, $u = \operatorname{Re} f$, where f is holomorphic. Consider the Cauchy integral formula

$$f(0) = \frac{1}{2\pi i} \int_\gamma \frac{f(t)}{t} \, dt \, .$$

This equality holds for any curve γ enclosing the point 0.

In the real-valued theorem we shall make explicit use of the fact that the curve γ is a circle. We parameterize a point of the circle as $t = Re^{i\varphi}$, so that $\dfrac{dt}{t} = \dfrac{iRe^{i\varphi} \, d\varphi}{Re^{i\varphi}} = i \, d\varphi$. When we substitute these values into the Cauchy integral, we find

$$f(0) = \frac{1}{2\pi} \int_0^{2\pi} f(t) \, d\varphi \, , \quad f = u + iv \, .$$

We obtain the equality we need separately for the real and imaginary parts. The theorem is now proved. □

In the multidimensional case the theories of harmonic and analytic functions diverge.

8.1. Consequences of the Mean-Value Theorem

1. The Maximum Principle. *A harmonic function has no extrema in the interior of its domain. More precisely, if $u(x_0, y_0) = \max u(x, y)$ and $(x_0, y_0) \in D$, then $u \equiv$ const.*

PROOF. Suppose the point (x_0, y_0) at which the maximum occurs is an interior point of the domain. Consider a small circle with center at that point, as in Fig. 8.3. If the value at some point (x_1, y_1) of the circle is less than at (x_0, y_0), then by continuity the function on a whole arc is less than at (x_0, y_0), violating the mean-value theorem. Hence the function must be constant in a neighborhood of the point (x_0, y_0). In other words, the set of points of the domain at which $u(x_0, y_0) = \max u(x, y)$ is open. But this set is obviously closed. Thus the set of interior maximum points is either empty or coincides with a component of the domain. As a result the function must be constant on the entire component of the domain containing (x_0, y_0). $\square$

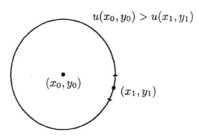

Fig. 8.3. The maximum principle

This proof relies only on the mean-value theorem. Thus the maximum principle will have been proved in the multi-dimensional case as soon as we prove the multidimensional mean-value theorem.

From the point of view of analytic function theory the maximum principle is a manifestation of the local open mapping principle. Indeed, if the real part of an analytic function could attain a maximum at an interior point, the open mapping principle would be violated, as shown in Fig. 8.4.

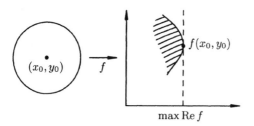

Fig. 8.4. The maximum principle and the open mapping principle

2. Uniqueness. *The continuous solution of the Dirichlet problem in a bounded connected domain is unique.*

PROOF. The difference of two solutions is a harmonic function in the interior of the region and identically equal to zero on the boundary. Since the domain is bounded, the maximum and minimum of that difference must be attained (either in the interior of the domain or on its boundary). If the maximum or minimum occurs in the interior, the difference is constant, and hence equal to zero. If it occurs on the boundary, then both the maximum and minimum equal zero, and so the difference is again identically zero. Thus by the maximum principle the difference is identically zero in the domain. □

We remark that the uniqueness of the solution of the Dirichlet problem holds for continuous functions satisfying the mean-value theorem, that is, there is at most one continuous function satisfying the mean-value theorem and having prescribed values on the boundary of a bounded domain.

Remark. In unbounded domains the solution is not unique. For example, the function x is harmonic in the half-plane $x > 0$, continuous on the closure of the half-plane, and identically equal to zero on the boundary, but not in the interior.

Theorem 3. *A continuous function satisfying the mean-value theorem is harmonic.*

PROOF. Consider the Dirichlet problem (in the ordinary sense, that is, for a harmonic function) in a domain with boundary values given by the values of a function satisfying the mean-value theorem. (We shall prove below that such a solution exists.) Consider the difference of the harmonic function and the function satisfying the mean-value theorem. It vanishes on the boundary and satisfies the mean-value theorem on the interior. Hence it must vanish identically. □

Remark. The fact that the kernel of the Laplacian consists of very smooth functions – the real parts of analytic functions – is inherent in the nature of this operator. For the wave operator (d'Alembertian) the picture is quite different. A singularity in the data for the Cauchy problem for the wave equation propagates along characteristics, so that the solutions of the wave equation cannot be smoother than the initial data.

We now turn to the solution of the Dirichlet problem. We first construct the solution for a disk.

OUTLINE OF THE FIRST METHOD. We assemble a rather large supply of harmonic functions in the disk. On the boundary of the disk we consider the function $\cos n\varphi$. Is there an analytic function with this real part? There is, namely,

$$f_n(z) = \left(\frac{z}{R}\right)^n = \left(\frac{|z|}{R}\right)^n (\cos n\varphi + \mathrm{i} \sin n\varphi) .$$

The real part of the function $\mathrm{i} f_n(z)$ on the boundary $|z| = R$ is $-\sin n\varphi$. Thus we can solve the Dirichlet problem if the boundary condition has the form $\cos n\varphi$ or $\sin n\varphi$. But we know that any boundary condition can be expanded in a series of these functions. Hence we can solve the problem with any boundary condition.

We now carry out the first method in the Dirichlet problem for the disk. We expand the boundary condition in a Fourier series:

$$f(\varphi) = \frac{a_0}{2} + \sum (a_n \cos n\varphi + b_n \sin n\varphi) \ .$$

Passing to polar coordinates, we write the harmonic extention of the functions $\cos n\varphi$ and $\sin n\varphi$ over the disk:

$$\left(\frac{r}{R}\right)^n \cos n\alpha \ , \quad \left(\frac{r}{R}\right)^n \sin n\alpha \ .$$

As a result we obtain the solution in the form of a series:

$$u(r,\alpha) = \frac{a_0}{2} + \sum \left(a_n \left(\frac{r}{R}\right)^n \cos n\alpha + b_n \left(\frac{r}{R}\right)^n \sin n\alpha\right) . \qquad (*)$$

Our aim is to represent the solution in integral form. The Fourier coefficients are

$$a_n = \frac{1}{\pi} \int_0^{2\pi} f(\beta) \cos n\beta \, d\beta \ ,$$

$$b_n = \frac{1}{\pi} \int_0^{2\pi} f(\beta) \sin n\beta \, d\beta \ ,$$

$$a_0 = \frac{1}{\pi} \int_0^{2\pi} f(\beta) \, d\beta.$$

We substitute these relations into $(*)$, obtaining

$$u(r,\alpha) = \frac{1}{\pi} \int_0^{2\pi} f(\beta) \left[\sum \left(\frac{r}{R}\right)^n \cos n(\alpha - \beta) + \frac{1}{2}\right] d\beta \ ,$$

where

$$z = (r,\alpha) \ , \quad \left(\frac{r}{R}\right)^n \cos n(\alpha - \beta) = \operatorname{Re}\left(\frac{z}{t}\right)^n \ , \quad t = Re^{i\beta} \ .$$

We now sum the geometric series:

$$\left(1 + \frac{z}{t} + \left(\frac{z}{t}\right)^2 + \cdots\right) - \frac{1}{2} = \frac{1}{1 - \frac{z}{t}} - \frac{1}{2} = \frac{t}{t-z} - \frac{1}{2} \ ,$$

obtaining

$$u(r,\alpha) = \int_0^{2\pi} f(\varphi) \operatorname{Re}\left(\frac{t+z}{t-z}\right) d\varphi \ .$$

We now transform the kernel: $\dfrac{t+z}{t-z} = \dfrac{(t+z)(\overline{t-z})}{(t-z)(\overline{t-z})}$.

The real part of the numerator is $t\bar{t} - z\bar{z} = R^2 - r^2$, since the remaining expression $t\bar{z} - \bar{t}z$ is purely imaginary.

The denominator $|t-z|^2$ equals $R^2 + r^2 - 2Rr\cos(\alpha - \beta)$.

Thus we obtain the *Poisson kernel*

$$\frac{R^2 - r^2}{R^2 + r^2 - 2Rr\cos(\alpha - \beta)} \,.$$

To summarize:

$$u(r,\alpha) = \frac{1}{\pi} \int\limits_{0}^{2\pi} \frac{R^2 - r^2}{R^2 + r^2 - 2Rr\cos(\alpha - \beta)} f(\beta)\, \mathrm{d}\beta \,.$$

Problem. Prove that this integral can be obtained by a conformal mapping of the solution we shall obtain below for the half-plane (second method).

By the Riemann mapping theorem any simply connected bounded domain can be conformally mapped onto a disk, and hence a solution of the Dirichlet problem can be obtained in any such domain using the Poisson kernel, provided the conformal mapping is known.

OUTLINE OF THE SECOND METHOD.

Problem. Let u be a solution of the Dirichlet problem in a domain D_1, and let $f : D_1 \to D_2$ be a conformal mapping of the domains. Construct a solution of the Dirichlet problem in the domain D_2. Try to convince yourself that this solution works without computing the Laplacian.

We know that a disk can be conformally mapped onto a half-plane. Let us try to solve the Dirichlet problem for a half-plane: $\Delta u = 0$, $u(x,0) = g(x)$.

We begin by noting that for the simple boundary condition shown in Fig. 8.5 the solution is the function $\frac{1}{\pi}\varphi_a(x,y)$, where

$$\varphi_a(x,y) = \operatorname{Im}\ln(z-a) = \arctan\frac{y}{x-a}\,.$$

Indeed, this function is harmonic in the half-plane and satisfies the boundary condition of Fig. 8.5. We now consider the "corner" function

$$F_{ab}(x,y) = \frac{1}{\pi}\bigl(\varphi_b(x,y) - \varphi_a(x,y)\bigr)$$

Fig. 8.5. A boundary condition
for the Dirichlet problem
in a half-plane

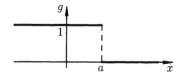

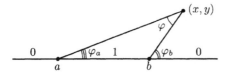

Fig. 8.6. The "corner" function

(see Fig. 8.6). It is harmonic in the half-plane and satisfies the boundary condition $g(x) = 1$ for $a < x < b$ and $g(x) = 0$ for $x < a$ and $x > b$.

We approximate an arbitrary boundary condition $g(x)$ (say, continuous and equal to zero outside a finite interval) by a piecewise-constant function, as in Fig. 8.7, and the corresponding harmonic function is $\sum \dfrac{F_{ab}g(a)(x - a)}{b - a}$.

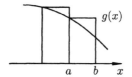

Fig. 8.7. Approximation by a piecewise-constant boundary condition

We now pass to the limit as the partition is refined, obtaining a harmonic function with boundary condition $g(x)$ written in the following integral form

$$\int \lim_{b \to a} \frac{F_{ab}(x, y)}{b - a} g(a) \, da .$$

The function $\lim_{b \to a} \dfrac{F_{ab}(x, y)}{b - a}$ in this integral is called the *kernel*. Let us study its level lines. The picture of the level lines for $F_{ab}(x, y)$ is shown in Fig. 8.8. A conformal mapping leaving the point 0 fixed and mapping a to ∞ maps this picture to the simple picture of the level lines for the corner. After dividing by the length of the interval and passing to the limit, we obtain the picture of the level lines shown in Fig. 8.9.

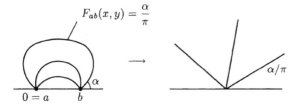

Fig. 8.8. Level lines for the function F_{ab}

Fig. 8.9. Level lines of the kernel

It is clear that some computations are required here: the derivative of the function $\frac{1}{\pi}\arctan\frac{y}{x-a}$ with respect to the parameter a is

$$\frac{d}{da}\left(\frac{1}{\pi}\arctan\frac{y}{x-a}\right) = \frac{1}{\pi}\frac{1}{1+\left(\dfrac{y}{x-a}\right)^2}\frac{y}{(x-a)^2} = \frac{1}{\pi}\frac{y}{(x-a)^2+y^2}\,.$$

Finally,

$$u = \frac{1}{\pi}\int\limits_{-\infty}^{+\infty}\frac{y}{(x-a)^2+y^2}g(a)\,\mathrm{d}a\,.$$

The picture of the level lines for the kernel shown in Fig. 8.10 is well-known from physics – it is the picture of the equipotential lines of a so-called *dipole* (the lines of force of a dipole, which are orthogonal to the equipotential lines, form the same kind of picture).

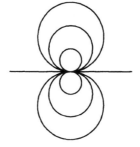

Fig. 8.10. A dipole

8.2. The Mean-Value Theorem in the Multidimensional Case

The Maximum Principle in the Multidimensional Case. We shall give a geometric explanation of the fact that a harmonic function cannot have a strict maximum at an interior point of its domain. A harmonic function minimizes the Dirichlet integral $\int_D(\nabla u)^2\,\mathrm{d}x$.

If there were a strict interior maximum, the graph would be bell-shaped in a neighborhood of that point, as shown in Fig. 8.11. Cutting off the top of the bell with a horizontal plane, we would obtain a function with a smaller Dirichlet integral. Thus a function with an interior extremum cannot minimize this integral. To be sure, this reasoning does not encompass functions with nonstrict maxima, as in Fig. 8.12. Nevertheless, we have proved that a function equal to zero on the boundary of a bounded domain and minimizing the Dirichlet integral vanishes identically on the domain.

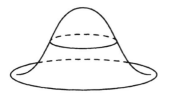

Fig. 8.11. A function that does not minimize the Dirichlet integral

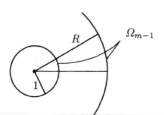

Fig. 8.12. A nonstrict maximum

The full maximum principle follows from the multidimensional mean-value theorem.

PROOF OF THE MEAN-VALUE THEOREM. On the sphere of radius R with center at 0 in $\mathbb{R}^m$ consider the element of solid angle Ω_{m-1}, that is, the projection of the element of surface area on the concentric sphere of radius 1, as shown in Fig. 8.13. We set $f(R) = \int\limits_{|x|=R} u(x)\,\Omega_{m-1}$. Then

$$f'(R) = \int\limits_{|x|=R} \frac{\partial u}{\partial n}\,\Omega_{m-1} = \int\limits_{|x|=R} (\nabla u, n)\,\mathrm{d}\Omega_{m-1}\,,$$

Fig. 8.13. On the mean-value theorem

and this is the flux of the gradient of u across the sphere divided by R^{m-1}. But if u is harmonic, that flux is 0. Thus $f'(R) = 0$, and so $f(r)$ is constant. But for small values of R it is obvious that this integral is approximately $u(0)$. Hence $f(R) = u(0)$, which was to be proved. $\square$

Corollary. $u(0) = \dfrac{1}{V_R} \int\limits_{|x| \le R} u(x)\,dx$, *where V_R is the volume of a ball of radius R.*

Indeed, to integrate over the ball we can integrate first over a sphere of fixed radius and then with respect to the radius.

Corollary. *The maximum principle holds for a harmonic function in n-dimensional space.*

Corollary. *The Dirichlet problem has a unique solution in the n-dimensional case.*

Remarks. 1. When deriving the Poisson formula we did not verify convergence out to the boundary. Of course our reasoning shows that if there exists a formula for the solution of the Dirichlet problem, it must have the form

$$u(r,\alpha) = \frac{1}{\pi} \int\limits_{0}^{2\pi} \frac{R^2 - r^2}{R^2 + r^2 - 2Rr\cos(\alpha - \beta)} f(\beta)\,d\beta\ . \qquad (**)$$

Nevertheless, it is still necessary to verify carefully that for every continuous function f the function u defined by formula $(**)$ indeed solves the Dirichlet problem. Try to carry out the necessary estimates independently. (In case of difficulty the reasoning can be found in the textbook of H. Cartan *Elementary Theory of Analytic Functions of One and Several Variables*, Chap. IV, §4, par. 4.)

2. Of course, the second derivation of the Poisson formula (for the half-plane) has the same gap as the first – we explain how to write the Poisson kernel, but we do not verify that the resulting solution works for every continuous function on the boundary.

3. In our second derivation of the Poisson formula we first solve the Dirichlet problem for very simple, but still discontinuous initial data. The solutions we found (connected with the "corner" function) are bounded. Only for bounded solutions can one guarantee the uniqueness of the solution of the Dirichlet problem with discontinuous boundary data, but we have not stated or proved this theorem. All this emphasizes the heuristic character of our derivation of the Poisson formula.

4. Here is an example of the nonuniqueness of the solution of the Dirichlet problem for discontinuous initial data. Consider the Dirichlet problem in the half-plane $y \ge 0$ with initial condition $u(x,0) \equiv 0$. There are solutions that are discontinuous at ∞. Consider any entire function $f(r)$ that is real-valued on the real axis (for example, $f(z) = z$ or $f(z) = \exp(z)$ and the like). Then the function $u = \operatorname{Im} f(z)$ gives a solution of our problem. We note that all these solutions, except $u \equiv 0$, are unbounded. Prove this using analytic function theory.

Prove also a more general fact: two bounded functions that have only a finite number of points of discontinuity in a closed disk, are harmonic on the interior of the disk, and coincide on the boundary coincide at all their points of continuity.

Lecture 9

The Fundamental Solution for the Laplacian.
Potentials

The interaction of mathematics and physics is proceeding along strange paths at present. Dirac, one of the greatest physicists of the twentieth century, stated the strategy of theoretical physics as follows: "In setting about to construct a physical theory, one must reject all previous physical models, as well as the 'physical intuition' based on them, which is nothing more than a collection of preconceived points of view." We must, according to Dirac, simply choose a beautiful mathematical theory and systematically find physical interpretations for its consequences, not fearing any contradiction with previous theories.

> I learnt to distrust all physical concepts as the basis for a theory. Instead, one should put one's trust in a mathematical scheme, even if the scheme does not appear at first sight to be connected with physics. One should concentrate on getting an interesting mathematics.

> *P. M. Dirac*

(Cited in the book *N. Wiener*, by P. Massani, Birkhäuser, 1990, p. 6.)

Amazing as it is, the whole theoretical physics of the twentieth century (in contrast to the physics of the nineteenth century) confirms Dirac's correctness. It has been remarked that every new physical theory refutes all preceding theories, while the mathematical models remain.

But today we shall take up a mathematical theory that, to the contrary, long remained unrecognized by mathematicians, although physicists had long made free use of it – the theory of so-called *generalized functions*. The most important example of a generalized function is the Dirac δ-function. We first study it on the line, then in $\mathbb{R}^n$; it could be defined on any manifold. This is the mathematical analog of such physical concepts as a point charge and a point mass.

The physical "definition," against which orthodox mathematicians protest, is as follows: the δ-function equals 0 everywhere except at zero, where it equals infinity. Its integral over the entire line is 1. Physicists have an excellent ability to work with such definitions, which have no explicit mathematical meaning.

We shall adopt the following definition, which is simple in application, makes it possible to explain the physical meaning easily, and is less tedious than the definition in the "correct mathematical theory of generalized functions."

Our δ-function of the variable x is denoted $\delta(x)$ and concentrated at 0. When $\delta(x)$ occurs in a formula, it is to be replaced by a δ-shaped function $\delta_\varepsilon(x)$, and the limit is to be taken as ε tends to zero.

A *δ-shaped function* is an ordinary smooth function with the following properties:

1. Its integral over the entire line equals 1.
2. It is nonnegative.
3. It is concentrated in the interval $(-\varepsilon, \varepsilon)$, that is, it vanishes outside this interval.

(These last two conditions could be weakened.)

The typical form of the graph of a δ-shaped function is shown in Fig. 9.1.

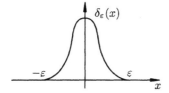

Fig. 9.1. A δ-shaped function

In passing to the limit one can fix the δ-shaped function. (The number ε occurs as a parameter in the formula.) One may also leave it unspecified, provided the family $\delta_\varepsilon(x)$ satisfies conditions 1)–3).

9.1. Examples and Properties

1. $\int \delta(x)\,dx = \lim\limits_{\varepsilon \to 0} \int \delta_\varepsilon(x)\,dx = 1$.

2. $\delta(x) = 0$ when $x \neq 0$, since $\lim\limits_{\varepsilon \to 0} \delta_\varepsilon(x) = 0$ for $x \neq 0$.

3. Let us compute $\int f(x)\delta(x)\,dx$, where $f(x)$ is a continuous function.

The main contribution to the integral $\int f(x)\delta_\varepsilon(x)\,dx$ is given by the expression $\int f(0)\delta_\varepsilon(x)\,dx$ in Fig. 9.2, which tends to $f(0)$ as $\varepsilon \to 0$; the difference between it and the whole integral also tends to zero as $\varepsilon \to 0$. (This is an exercise.) As a result, $\int f(x)\delta(x)\,dx = f(0)$.

4. Similarly, $\int f(y)\delta(x - y)\,dy = f(x)$.

The reasoning is the same, taking into account that $\delta_\varepsilon(x-y)$ is the translate of the function $\delta_\varepsilon(x)$ from a neighborhood of 0 to a neighborhood of y, as in Fig. 9.2.

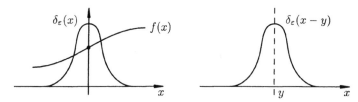

Fig. 9.2. Computing the convolution with the δ-function

The formula of item 4 also has the following remarkable interpretation: an arbitrary function f is a continuous linear combination of translated δ-functions $\delta(\cdot - y)$ concentrated at the points y with coefficients $f(y)$, so that the δ-functions $\delta(\cdot - y)$ form a "continuum size basis" in the space of functions on the line. It is also useful to interpret $\delta(\cdot - y)$ as the density of a unit point mass or charge concentrated at y.

5. *Homogeneity.*

Definition. A function f is *homogeneous of degree* d if $f(\lambda x) = \lambda^d f(x)$ for any positive λ.

Let us find out whether the δ-function is homogeneous, and if so, of what degree. As an example, let us find $\delta(2x)$. Substitute $2x$ in place of x in the function $\delta_\varepsilon(x)$. The value of the integral is then cut in half. Passing to the limit, we find $\delta(2x) = \frac{1}{2}\delta(x)$. The reasoning is analogous for any multiple. Thus, the function $\delta(x)$ is homogeneous of degree -1, just like the function $1/x$. This similarity is not accidental. In a certain sense these functions are related, to be sure, only in the one-dimensional case.

In the n-dimensional case the function $\delta(x)$ is homogeneous of degree $-n$. One can also verify this as follows: $\delta(x_1, \ldots, x_n) = \delta(x_1) \cdots \delta(x_n)$, since the corresponding product of δ-shaped functions gives a δ-shaped function in n-dimensional space. Since the factors are homogeneous of degree -1, the product is obviously homogeneous of degree $-n$.

9.2. A Digression. The Principle of Superposition

Although this principle is the basis of all of so-called linear physics, it is in essence a simple fact of linear algebra.

Consider a linear transformation $L : V \to W$ mapping one vector space into another, and consider a nonhomogeneous linear equation $Lu = f$. The equation has a solution if and only if $f \in \mathrm{Im}\, L$, where $\mathrm{Im}\, L$, the image of L, is a subspace of W. The solutions form an affine subspace in V parallel to $\mathrm{Ker}\, L$, where $\mathrm{Ker}\, L$, the kernel of L, is a subspace of V. To be specific, let u_p be a particular solution of the nonhomogeneous equation. Its general solution can then be represented as $u = u_\mathrm{p} + u_\mathrm{h}$, where u_h is the general solution of the homogeneous equation $Lu = 0$, as shown in Fig. 9.3.

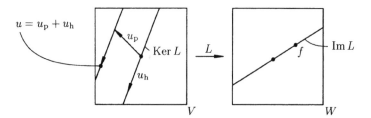

Fig. 9.3. Solutions of a nonhomogeneous linear equation

We now state the principle of superposition itself. Let u_1 and u_2 be respectively solutions of the equations $Lu = f_1$, $Lu = f_2$. Then $\alpha_1 u_1 + \alpha_2 u_2$ is a solution of the equation $Lu = \alpha_1 f_1 + \alpha_2 f_2$. Here α_1 and α_2 are arbitrary constants.

Physically, if we throw one stone into water we obtain a certain picture of the waves; if we throw a second stone, we obtain a second picture of the waves. If we throw both stones together, the resulting picture of the waves is the same as if we added the disturbances from throwing the first stone to those from throwing the second.

The proofs of all the assertions listed above are obvious by the linearity of the operator.

We now combine the principle of superposition with the possibility of expanding any function into a superposition of δ-functions. We obtain the following. If we want to solve a linear differential equation with an arbitrary function f on the right-hand side, it suffices to learn how to solve this equation with the δ-function on the right, then apply the principle of superposition, except that instead of sums one must write integrals.

Let us apply these considerations to the *Poisson equation* $\Delta u = f$. We must first solve the equation $\Delta u = \delta(x)$. Of course, the solution of this equation is not unique. Indeed, the kernel of Δ consists of all harmonic functions, so that an arbitrary harmonic function can be added to the solution.

Of all the solutions of the equation $\Delta u = \delta(x)$ one can exhibit one remarkable one called the *fundamental* solution. Let us find the grounds on which it can be exhibited.

The Laplacian is invariant with respect to the group of Euclidean motions (rotations and translations). Indeed, the Laplacian is the divergence of the gradient and hence is defined by the Euclidean structure alone. On the other hand, the δ-function is also spherically symmetric. Therefore a solution of this equation is again obtained by rotating any solution of $\Delta u = \delta$ around the origin. The arithmetic mean of two solutions is also a solution. In general any linear combination of several solutions whose coefficients add to 1 is again a solution. For that reason, one can average a solution over the group of all rotations about the origin, which is a compact group.

For example, when $n = 2$, the group $SO(2)$ of rotations of the plane is isomorphic to the circle S^1, and the averaging measure is $\frac{1}{2\pi}\,d\varphi$.

When $n = 3$, the group of rotations $SO(3)$ is isomorphic to the real three-dimensional projective space $\mathbb{R}P^3$. It has a two-sheeted covering by the three-dimensional sphere (the group of unit quaternions), which in turn is isomorphic to the special unitary group $SU(2)$, also known as the spin group of order 3, as in the following diagram:

$$S^3 \cong SU(2) = \mathrm{Spin}\,3$$

$$2\Big\downarrow$$

$$SO(3) \cong \mathbb{R}P^3$$

In particular, a metric can be carried over from S^3 to $SO(3)$ under the two-sheeted covering. This metric is rotation-invariant, so that there is a measure with which one can carry out averaging.

After averaging we obtain a solution that is rotation invariant. In other words, the value of the solution $u(x)$ depends only on the distance from x to the origin. For such functions the operator Δ becomes an ordinary second-order differential operator. For that reason the equation for u has the form $u'' + A(r)u' + B(r)u = 0$ for $r > 0$. (Here the right-hand side is zero, since the δ-function vanishes for $r > 0$.)

The coefficients A and B can be computed explicitly, if one does not mind passing to polar coordinates in the Laplacian, but one can also get by without any computations. First of all, we know the obvious solution $u \equiv 1$. Hence $B \equiv 0$, so that the operator has the form $u'' + A(r)u'$.

We now apply homogeneity considerations. The Laplacian maps homogeneous functions into homogeneous functions of degree two less. Since $\delta(x)$ is homogeneous of degree $-n$, a homogeneous solution u of the equation $\Delta u = \delta$ must be homogeneous of degree $2 - n$. Consequently it must be proportional to $1/r^{n-2}$. In particular, when $n = 3$, it is proportional to $1/r$. (We remark that when $n = 2$ this formula does not work – there is no homogeneous non-constant solution in this case.)

Let us now compute the Laplacian: $\Delta(1/r^{n-2}) = \operatorname{div}\operatorname{grad}(1/r^{n-2})$.

The gradient of our homogeneous function is a vector field directed toward the center along radii; its length equals the absolute value of the derivative with respect to r:

$$(r^{2-n})' = (2-n)r^{1-n} = \frac{2-n}{r^{n-1}} \ .$$

(To be verified: when $n = 3$, the gradient of the potential $1/r$ is a Newtonian force, inversely proportional to r^2.)

To compute the divergence we remark that it is rotation invariant. We compute the flux of the gradient field across two concentric spheres, as in Fig. 9.4.

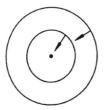

Fig. 9.4. Computation of div grad r^{2-n}

The flux across a sphere equals the magnitude of the field multiplied by the surface area of the sphere. The area of a sphere of radius R is $\omega_{n-1}R^{n-1}$, where ω_{n-1} is the area of the unit sphere (4π for $n = 3$, 2π for $n = 2$). As a result the flux is $(2 - n)\omega_{n-1}$. In particular, for $n = 3$ it is -4π.

Our field thus has the remarkable property of "incompressibility": its flux is the same across any sphere centered at 0. A spherically symmetric field is unambiguously determined by this condition (up to a constant factor). For example, in three-dimensional space only a Newtonian field has this incompressibility property. On the plane there also exists an $O(2)$-invariant incompressible field: the length of a vector of this field is inversely proportional to the distance from the center. This is a gradient field, but its potential is not a homogeneous function; it is $\ln \frac{1}{r}$. We thus arrive at the remarkable formula $\frac{1}{r^0} \cong \ln \frac{1}{r}$.

Remark. This formula could have been guessed, and along with it the fundamental solution of the equation $\Delta u = \delta$ in the plane, by considering the dimension of the plane $n = 2$ as the limit of the dimensions $n = 2 + \varepsilon$:

$$\frac{1}{r^{n-2}} = \frac{1}{r^\varepsilon} = \exp\left(\varepsilon \ln \frac{1}{r}\right) = 1 + \varepsilon \ln \frac{1}{r} + O(\varepsilon^2) \ .$$

Let us return to the computation of the divergence. The total flux of the field across the boundary of the layer between concentric spheres is zero. But this flux is the integral of the divergence over the layer. Thus

$$\int_{r_1}^{r_2} r^{n-1} \operatorname{div} \operatorname{grad}(r^{2-n}) \, dr = 0$$

for any r_1 and r_2, so that the divergence itself must be zero. Finally,

$$\Delta\left(\frac{1}{r^{n-2}}\right) = \operatorname{div} \operatorname{grad}\left(\frac{1}{r^{n-2}}\right) = 0$$

for $r > 0$. But then one can also find the coefficient $A(r)$. Thus we have avoided a lengthy change of variables.

Claim. $\Delta(1/r^{n-2}) \neq 0$ *throughout space.*

Indeed, consider the flux of the gradient field across a sphere with center 0, which equals $(2 - n)\omega_{n-1}$. Moreover, the sphere can be replaced by the boundary of an arbitrary region. If this region contains 0, the integral of the divergence over the region equals $(2 - n)\omega_{n-1}$, while the integral is 0 if the region does not contain 0. But this means, by the very definition of the δ-function, that the divergence of the field equals

$$(2 - n)\omega_{n-1}\delta(x) .$$

For example, when $n = 3$, we have $\Delta(1/r) = -4\pi\delta(x)$, and so forth.

Of course this result can also be verified using δ-shaped functions. But we see how much simpler it is to work directly with the ideal limiting object, the δ-function itself.

When $n = 2$, the function $\ln(1/r)$ will serve as the fundamental solution (potential). Indeed, the derivative of this function with respect to r is $-1/r$, and from that point on all our reasoning goes through.

We remark that on the level of fields the result is the same in all dimensions: the field is inversely proportional to r^{n-1}. The case $n = 1$ can also be included in the general picture: the fundamental solution of the operator d^2/dx^2 is the function $|tx|$ with a suitable t, and the magnitude of the field is proportional to $1/r^{1-1}$, that is, it is constant, as in Fig. 9.5.

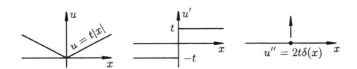

Fig. 9.5. The fundamental solution for the Laplacian when $n = 1$

We see that $t = 1/2$, so that the fundamental solution is $|x|/2$. It is interesting that the case $n = 1$ fits into the general picture even on the level of coefficients. To be specific, the general formula for a fundamental solution, which is

$$u_0 = \frac{1}{(2 - n)\omega_{n-1}r^{n-2}} ,$$

becomes $r/\omega_0 = r/2$ when $n = 1$. Indeed, the surface area (zero-dimensional volume) ω_0 of the zero-dimensional sphere S^0, which consists of two points, is 2.

Thus, the general formula for the fundamental solution holds in all dimensions except 2, where the logarithm appears: $u_0 = -\frac{1}{2\pi}\ln\frac{1}{r}$.

Remark. The appearance of the logarithm here has its own profound causes: the logarithm is a so-called *associated homogeneous function*. The logarithm is associated with power functions in exactly the same way as the adjoined vectors are associated with eigenvectors in the theory of Jordan cells. You have already encountered this phenomenon in the theory of resonance when solving homogeneous linear equations with constant coefficients: when the eigenvalues become multiple, quasi-polynomials are adjoined to the exponentials.

The physical meaning of the fundamental solution is the potential of a unit point charge at the point 0. The gradient of this potential is the field created by this charge.

It is physically interesting to consider fields with $n = 3$. But fields in smaller dimensions can also be obtained from three-dimensional pictures having suitable symmetry.

Consider, for example, a uniformly charged line in $\mathbb{R}^3$. Let us find the force created by this charge at a given point as the resultant of point charges smeared over the line. The result will be the same as in consideration of a plane passing through our point perpendicular to the charged line, and the whole charge must be placed at 0, as in Fig. 9.6.

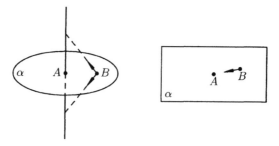

Fig. 9.6. A line generates the same field of charges in each plane orthogonal to it that a point charge generates

WARNING. It is important to add forces, not potentials, since the corresponding integral for the potentials diverges. We obtain another remarkable and amazing formula:

$$\int_{-\infty}^{+\infty} \frac{dz}{\sqrt{r^2 + z^2}} \sim 2 \ln \frac{1}{r} + \text{const} .$$

(It is forbidden ask the value of the constant!)

Similarly, if the charge is uniformly distributed on a plane, we obtain the same picture of forces on each line perpendicular to the plane that we would obtain from a point charge when $n = 1$.

A similar reduction can be carried out starting with any dimension.

We now consider a continuous function $f(x)$ defined in a bounded domain Ω. The solution u of the Poisson equation $\Delta u = f$ can be found by the principle of superposition, representing f as a superposition of translates of the δ-function. Let u_0 be the fundamental solution, that is $\Delta u_0(x) = \delta(x)$. In other words, $u_0(x)$ is the potential created at the point x by a unit charge located at the origin. If a charge of magnitude $f(y)$ is located at y, it creates the potential $f(y)u_0(x - y)$ at x.

The superposition of all these fields is the field with potential given by the formula

$$u(x) = \int_\Omega f(y)u_0(x - y)\,dy\ .$$

This function is called the *Poisson integral*. By the principle of superposition

$$\Delta u(x) = \int_\Omega f(y)\delta(x - y)\,dy = f(x)\ .$$

Therefore $u(x)$ is a solution of Poisson's equation.

Theorem. *Let the continuous function $f(x)$ have support in a bounded domain Ω. Then there exists a solution of the Poisson equation $\Delta u = f$ given by the formula*

$$u(x) = \int_\Omega f(y)u_0(x - y)\,dy\ ,$$

where u_0 is the fundamental solution for the Laplacian.

Once an explicit formula for the solution is written, it can be verified directly that the equation is satisfied, without resorting to δ-functions.

Exercise. Prove the theorem directly.

The solution is not unique, since any harmonic function can be added to it. A unique solution can be exhibited by imposing a condition at infinity.

When $n > 2$ the fundamental solution vanishes at infinity, and hence the entire Poisson integral vanishes. This condition identifies the solution uniquely. Indeed, the difference of two such solutions would be a harmonic function vanishing at infinity. By the maximum principle, it would then have to be identically zero.

When $n = 2$, a solution that is unique up to an additive constant can be singled out by the condition $|u(x)| \leq C \ln |x|$.

Now let us consider charges concentrated only on the boundary $\partial\Omega$ of the domain Ω. We denote the surface charge density at a point y of the boundary by $f(y)$. We set

$$u(x) = \int_{\partial\Omega} f(y)u_0(x - y)\,dy\ .$$

This function is harmonic everywhere outside the boundary of the domain Ω and vanishes at infinity. (When $n = 2$, $|u|$ grows no faster than the logarithm.)

What happens at points of the boundary itself? Consider an element dS of the boundary. We construct a cylinder over dS along the normal with small height ε, as in Fig. 9.7.

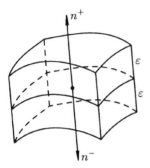

Fig. 9.7. The field created by a single layer

Consider the flux of the gradient of u across the boundary of the region G just created. One would naturally expect the flux across the lateral surface to be small as ε is, since this lateral surface has small area. (We leave this point unproved for the present.) We first compute the total flux:

$$\int_{\partial G} \frac{\partial u}{\partial n}\, d\sigma = \int_{\partial G} (n, \nabla u)\, d\sigma = \int_G \Delta u\, dx = \int_{(\partial \Omega) \cap G} f\, dS .$$

(The second of these equalities uses Stokes' theorem; the third takes account of the charges in G.) On the other hand, the flux across the base is $\left(\frac{\partial u}{\partial n_+} + \frac{\partial u}{\partial n_-} \right) dS$ up to a correction that is small compared with dS. Neglecting the flux through the lateral surface, which is small when ε is small, we conclude that the equality $\frac{\partial u}{\partial n_+} + \frac{\partial u}{\partial n_-} = f$ holds at each point of $\partial \Omega$.

The sum on the left is the *jump* in the normal derivative if we regard the normal as oriented in one direction. Thus, *the function $u(x)$ is harmonic in the interior and exterior of the surface, and the jump of its normal derivative equals the density of the charge distributed on the surface.* This function is called a *single-layer potential* – the charges are distributed in a single layer over the surface.

Example 1. Let us compute the potential of a sphere with a constant charge density. Up to a constant factor we obtain $u(x) = \int_{|y|=R} \frac{dy}{|x-y|}$, as in Fig. 9.8. This integral over the sphere can be computed explicitly, although the computation is complicated. (Newton was able to handle this.)

But the answer can also be found without integration, using the symmetry of the sphere. We know in advance that $u = u(r)$ is a harmonic function inside the sphere and outside it. Hence we can seek a solution in the form $a + b/r$.

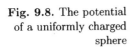

Fig. 9.8. The potential
of a uniformly charged
sphere

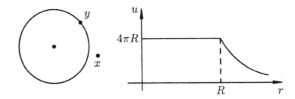

(The constants can be different inside and outside the sphere.) Inside the
sphere $b = 0$, since if not, the function would not be harmonic at 0. Thus u
is constant inside the sphere, and therefore equal to its value at the center,
namely $-4\pi R$. The force field inside the sphere is therefore zero.

As it happens, Newton discovered this by balancing forces, as shown in
Fig. 9.9. His reasoning is also valid for an ellipsoid. More precisely, for an
infinitely thin homogeneous layer between similar ellipsoids with a common
center, where symmetry considerations (the dependence of the potential only
on the radius) do not carry through.

Fig. 9.9. There is no attraction
inside a sphere or an ellipsoid

Problem. Prove that the attractive forces of opposite elements of the layer
between concentric similar ellipsoids are negatives of each other.

Let us return to our sphere and find the constants a and b for the exterior
region. On the outside $a = 0$ (the potential is zero at infinity). The coefficient
b is the total charge and equals $4\pi R$. One can also find a from consideration
of the continuity of the function u at the points of the sphere itself under
approach from within and without, since the integral converges uniformly.
The graph of the potential is shown in Fig. 9.8.

Example 2. Let A be a symmetric operator in Euclidean space and I the
identity operator. The operator $A - \lambda I$ defines a one-parameter family of
quadratic forms $((A - \lambda I)x, x)$, called a Euclidean bundle. The operator
$(A - \lambda I)^{-1}$ (the resolvent) is also symmetric and defines the nonlinear bundle
of quadratic forms $((A - \lambda I)^{-1}x, x)$ (dual to the Euclidean bundle in the sense
of projective duality). Consider the family of quadrics $((A - \lambda I)^{-1}x, x) = 1$.
For example, when $n = 2$, the bundle has the form of Fig. 9.10. All the curves
here have the same foci.

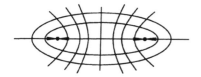

Fig. 9.10. The curves of the bundle for $n = 2$

The equations of the quadrics of the bundle can be written in an eigenbasis of the operator A as

$$\frac{x_1^2}{a_1 - \lambda} + \cdots + \frac{x_n^2}{a_n - \lambda} = 1 ,$$

where $a_1, \ldots, a_n$ are the eigenvalues of the operator A. Quadrics corresponding different values of λ are said to be *confocal* to each other.

Problem. Prove that precisely n mutually confocal quadrics of the bundle pass through each point of n-dimensional space (corresponding to some values $\lambda_1, \ldots, \lambda_n$). These quadrics are pairwise orthogonal at each point of their intersection.

The values $\lambda_1, \ldots, \lambda_n$ are called the *elliptic coordinates* of the point.

When $n = 3$, an ellipsoid, a hyperboloid of one sheet and a hyperboloid of two sheets pass through each point of space, as shown in Fig. 9.11.

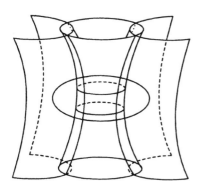

Fig. 9.11. The surfaces of the bundle for $n = 3$

Problem. Consider an ellipsoid and an infinitely near concentric similar ellipsoid. Consider the potential created by a uniform distribution of charge in the layer between these two ellipsoids. Such a potential is given by the density $\dfrac{dx_1 \wedge \cdots \wedge dx_n}{df}$, that is, $\dfrac{ds}{|\nabla f|}$, where ds is the Euclidean area and f the quadratic form that gives the ellipsoid. It is called a *homeoid* density. Then

1. The electrostatic field inside the ellipsoid is zero.

2. On the outside the lines of force (phase lines of the vector field given by the gradient of the potential) are coordinate lines of the elliptic coordinates described above.
3. All the level surfaces of the potential in the exterior are ellipsoids confocal with one another.

We recall some simple propositions of electrostatics.

The potential of a steady charge distribution of a conducting surface is constant on the surface. Indeed, if the potential were not constant the potential gradient (that is, the electrostatic field) would have a component in the direction of the surface, which would cause the charges to move.

The homeoid distribution just defined is the charge distribution on the surface of a conducting ("metallic") charged ellipsoid.

It is interesting that the charge density on a conducting surface is greater where the curvature of the surface is greater. Inside a conductor there cannot be any charges of a steady distribution. Indeed, the gradient of the potential of a steady charge distribution equals zero inside a conductor (otherwise the charges would begin to move). Hence the potential inside a conductor is constant and consequently the charge density there is zero.

In exactly the same way the potential of a smooth conducting surface bounding a bounded domain is constant in that domain if there is no charge.

Indeed, the potential is harmonic inside the surface and constant on the surface. Consequently, by the maximum principle it is also constant inside the domain.

Simply put, an electrostatic field cannot penetrate to the interior of a metallic cavity ("screening"). Analogous facts hold in $\mathbb{R}^n$ for any n.

Problem 1. A unit charge is placed at a point at distance r from the center of a noncharged conducting circle of radius R in $\mathbb{R}^2$. Find the electrostatic field created, its equipotential lines, and its lines of force.

HINT. Start with the case $r = 0$.

Problem 2. Carry out the computations for the density of an equilibrium distribution of charges for a "charged" conducting ellipse or square in the two-dimensional problem. What is the singularity of the charge distribution at the vertices?

9.3. Appendix. An Estimate of the Single-Layer Potential

We shall prove that the electrostatic field created by a smooth single-layer charge distribution on a smooth bounded surface is bounded all the way to the boundary.

Hence it follows in particular that the flux across the lateral surface of a cylinder of small height ε (in the constructions on p. 86) is small compared with ε, a fact we made use of in proving the formula for the jump of the normal derivative of the single-layer potential.

Consider the family of normals to the surface of the layer. A sufficiently small neighborhood of this surface is fibered into segments of these normals. It suffices to prove that the force is uniformly bounded inside such an r-neighborhood (since outside the neighborhood the force is majorized by $\rho S/r^2$, where ρ is the maximal charge density and S is the area of the surface). Consider a segment of the normal passing through a point P on the surface. In a neighborhood of the point P we introduce Cartesian coordinates with origin at P: z along the normal to the surface and $\mathbf{x}$ in the tangent plane. If the radius r of the neighborhood of P is sufficiently small, the equation of the surface of the layer in our coordinates can be written as $z = h(\mathbf{x})$, $|h(\mathbf{x})| \leq C|\mathbf{x}|^2$.

To prove that the force is bounded it suffices to obtain a uniform estimate of just the force created by charges in the sphere of radius r just constructed. (The force due to other charges, say in the $r/2$-neighborhood of P, does not exceed $4\rho S/r^2$.)

A real difficulty occurs in the estimate of the force created near the point P by charges in the neighborhood. Here the boundedness is obtained only because opposite parts of the neighborhood pull in opposite directions.

We begin by considering the special case when the layer in our sphere is a horizontal plane ($z = 0$) and the charge density is constant.

In this case the horizontal components of the forces created on the z-axis by charges at opposite points $\mathbf{x}$ balance each other precisely. The resulting force is vertical and is given at the point Z as an integral (when the density is one):

$$F = \int_0^r \frac{Z\, 2\pi x\, dx}{(x^2 + Z^2)^{3/2}} = 2\pi \int_0^{r/Z} \frac{\xi\, d\xi}{(\xi^2 + 1)^{3/2}} = 2\pi \left(1 - \frac{Z}{\sqrt{r^2 + Z^2}} \right),$$

where $x = |\mathbf{x}|$. (For simplicity we assume Z is positive; the force does not change when Z changes sign.)

In any case the force is bounded by 2π (corresponding, as we already know, to the field of a uniformly charged plane).

We now replace the charge distribution along the surface $z = h(\mathbf{x})$ with density $\rho(\mathbf{x})$ in the region $x < r$ by a distribution of constant density $\rho(0)$ in the disk $x < r$ on the plane $Z = 0$. (Here Z is the coordinate of the point on the z-axis where we wish to estimate the magnitude of the field.)

The vertical component of the force at Z is now given by the integral of

$$\frac{(Z + h(\mathbf{x}))\rho(\mathbf{x})J(\mathbf{x})x\, dx\, d\varphi}{(x^2 + (Z + h)^2)^{3/2}},$$

where ρ is the density and J is the Jacobian, which is equal to the ratio of the element of area of the layer over $\mathbf{x}$ to the element of area $x\,dx\,d\varphi$ in the plane of $\mathbf{x}$ (with polar coordinates x, φ).

The quantities ρ and J are bounded in this region. In addition, the expression in the denominator is

$$x^2 + (Z + h)^2 = x^2 + Z^2 + 2Zh(\mathbf{x}) + h^2(\mathbf{x}) \geq \frac{1}{2}(x^2 + Z^2)$$

for sufficiently small r, since $|Zh| \leq \frac{Z^2}{4} + h^2$ and $|h| \leq Cx^2$.

For that reason the entire term proportional to Z in the integral can be estimated from above by a constant, just as in the case of a homogeneous plane.

The second term (proportional to h) in the integrand is bounded. (Its numerator is less than $C_1 x^3$ and its denominator is greater than $C_2 x^3$.) Thus the entire vertical component is bounded. The horizontal component, that is, the integral of $\dfrac{\mathbf{x}\rho(\mathbf{x})J(\mathbf{x})x\,dx\,d\varphi}{(x^2 + (Z + h(\mathbf{x}))^2)^{3/2}}$, is more difficult to estimate.

First of all, we may assume that $\rho J \equiv 1$. Indeed, the functions ρ and J are smooth, so that the replacement of $\rho(\mathbf{x})J(\mathbf{x})$ by $\rho(0)J(0)$ changes the integrand by a bounded term ($\leq C_3 x^3$ in the numerator and $\geq C_2 x^3$ in the denominator).

It remains to compare the integral for $\rho J = 1$ with the same integral when $h = 0$ (which is equal to zero, as we know).

Lemma.

$$\left| \frac{1}{(x^2 + Z^2)^{3/2}} - \frac{1}{(x^2 + (Z + h)^2)^{3/2}} \right| \leq \frac{C_4}{x^2}.$$

PROOF. Let $Z = \lambda x$. Then the left-hand side is

$$\frac{1}{x^3} \left| \frac{1}{(1 + \lambda^2)^{3/2}} - \frac{1}{(1 + (\lambda + \mu)^2)^{3/2}} \right|,$$

where $h = \mu x$, so that $|\mu| < Cx$.

The function $\dfrac{1}{(1 + \lambda^2)^{3/2}}$ has a derivative that is bounded by some constant C_5 on the entire line. By the formula for finite increments

$$\left| \frac{1}{(1 + \lambda^2)^{3/2}} - \frac{1}{(1 + (\lambda + \mu)^2)^{3/2}} \right| \leq C_5\mu.$$

Since $|\mu| \leq Cx$, the quantity being estimated in the lemma does not exceed

$$\left| \frac{C_5\mu}{x^3} \right| \leq \frac{C_5 C}{x^2},$$

which was to be proved. $\square$

Finally, for the integral that gives the horizontal component of the force when $J\rho \equiv 1$, we obtain the estimate

$$\left| \int_0^r \int_0^{2\pi} \frac{\mathbf{x}\, x\, dx\, d\varphi}{(x^2 + (Z + h(\mathbf{x}))^2)^{3/2}} \right| \le 2\pi \int_0^r \frac{x^2 C_4}{x^2}\, dx = 2\pi r C_4 \;.$$

(The same integral equals zero when $h = 0$, and the difference of the integrands for $h = 0$ and for any h is estimated by the quantity C_4/x^2 in the lemma.)

Remark. By similar elementary estimates one can also prove that the single-layer potential is continuous in the entire space and that the limits of its derivatives along exterior and interior normals exist and are continuous, nor merely the formula for the jump in the normal derivative to which we have confined ourselves here.

Remark. In the study of potentials there is a very useful convention due to Faraday that "4π electrostatic lines of force emanate from each unit charge" (whatever that may mean). It is assumed that the lines begin in positive charges (or at infinity) and end at negative charges (or at infinity).

In this case the field intensity turns out to be equal to the "density of the lines of force." For example, for single-layer charges half of the $\rho\, 4\pi\, dS$ lines of force emanating from the charges of the region dS go in one direction and the other half in the opposite direction, from which the formula for the jump follows. The factor 4π occurs here because physics considers the potential of a unit charge of the form $u = 1/r$, while we have defined the single-layer potential as the integral of translates of the fundamental solution of $\Delta u = \delta$.

Lecture 10

The Double-Layer Potential

We have thus found the fundamental solution for the Laplacian in $\mathbb{R}^n$:

for $n > 2$, $u_0 = \dfrac{1}{-(n-2)\omega_{n-1}r^{n-2}}$, where ω_{n-1} is the surface of the unit sphere in $\mathbb{R}^n$;

for $n = 2$, $u_0 = -\dfrac{1}{2\pi}\ln\dfrac{1}{r}$;

for $n = 1$, $u_0 = \dfrac{|x|}{2}$.

Incidentally the signs are determined by the condition $\Delta u_0 = \delta$ as follows: when the singularities near zero are smoothed out, the smoothed function must have a *positive* second derivative. The coefficient is determined from the condition that the flux of $\operatorname{grad} u$ across the unit sphere must be equal to 1.

Using fundamental solutions we constructed the following potentials: the solid potential, which gives a solution of the Poisson equation, and also the single-layer potential. We now consider a double-layer potential.

We consider a charge distribution consisting of two layers on a hypersurface, one layer positive and the other negative, with a certain density $\rho(x)$ (ρ is a function on the hypersurface), as in Fig. 10.1.

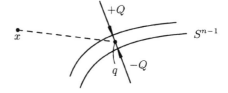

Fig. 10.1. A double layer on a hypersurface

Here we have in mind the following limiting passage. Let l be the distance between charges along the normal drawn to the hypersurface at the point q; the charges $+Q = \rho/l$ and $-Q = -\rho/l$ have magnitude of order $1/l$; they create a certain field. We compute the limit of the field as $l \to 0$. This limit is called the *dipole field*, and its potential is the *dipole potential*. The *dipole moment* $\rho = Ql$ is conserved in the limiting passage. The direction joining the infinitely near charges that make up the dipole is called the *dipole axis*. The potential of a double layer of density ρ distributed on a hypersurface is the integral of the dipole potentials of the charges located on the hypersurface whose axes are the normals to the hypersurface, the dipole moment in an infinitely small region ds on the hypersurface being $\rho\, ds$.

We now carry out the computation of the dipole potential in terms of the fundamental solution u_0:

$$u_0\left(x - \left(q + n_q \frac{l}{2}\right)\right)\frac{\rho}{l} - u_0\left(x - \left(q - n_q \frac{l}{2}\right)\right)\frac{\rho}{l}$$
$$= \rho\frac{\partial u_0(x - q)}{\partial n_q} + o(1) \quad \text{as } l \to 0 .$$

Here n_q is the outward normal to the hypersurface at q.

We have found the element of the double-layer potential formed by a dipole located at q; altogether the double-layer potential at x is

$$u(x) = \int_{S^{n-1}} \rho(q)\frac{\partial u_0(x - q)}{\partial n_q}\, dq .$$

10.1. Properties of the Double-Layer Potential

1. Geometric meaning of the double-layer potential. Instead of the fundamental solution normalized by the condition $\Delta u_0 = \delta$, it is customary to use the function $u_0 = 1/r^{n-2}$ (when $n > 2$) and $u_0 = \ln(1/r)$ when $n = 2$. The double-layer potential is a harmonic function in both the exterior and interior regions bounded by the surface, since this was true for the limiting passage

$$\Delta u\big|_{\mathbb{R}^n \setminus S^{n-1}} = 0 .$$

There may even be several exterior regions if the surface is not connected.

Example. Let $n = 2$ and $\rho \equiv 1$ on a connected closed curve. Then $u(x) = $ const, the constant being different outside, inside, and on the curve.

We shall now prove this remarkable fact. We first compute the dipole potential. The center of the dipole can be taken to be at the origin. The derivative of the fundamental solution in the direction of any vector **v** is

$$L_{\mathbf{v}} \ln \frac{1}{r} = -\frac{1}{r}L_{\mathbf{v}}r = -\frac{|\mathbf{v}| \cos \varphi}{r} .$$

Indeed, $L_{\mathbf{v}}r$ can easily be computed for any dimension:

$$L_{\mathbf{v}}\sqrt{x_1^2 + \cdots + x_n^2} = \frac{2(x_1 v_1 + \cdots + x_n v_n)}{2\sqrt{x_1^2 + \cdots + x_n^2}} = \frac{(\mathbf{v}, \mathbf{r})}{r} = v \cos\varphi \, ,$$

as shown in Fig. 10.2.

Fig. 10.2. The derivative of the radius-vector
with respect to the vector **v**

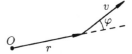

The derivative of the fundamental solution for any $n > 2$ is computed similarly:

$$L_{\mathbf{v}}\frac{1}{r^{n-2}} = \frac{(2-n)(\mathbf{r}, \mathbf{v})}{r^n} \, .$$

We remark that the potential of an isolated charge decreases at infinity like $1/r^{n-2}$, while that of a dipole decreases faster, like $1/r^{n-1}$. (The positive and negative charges are said to "screen" each other.)

Let us draw the level lines of the dipole potential (for $n = 2$). If we direct the x-axis along the vector **v**, we obtain the function $\dfrac{x}{r^2} = \dfrac{x}{x^2 + y^2}$. This is a very important harmonic function in the plane with the origin removed. It has a singularity at 0. Its level lines are conic sections given by the equations $x^2 + y^2 = cx$. They form a family of circles passing through the origin and tangent to one another at the origin. These circles are perpendicular to the axis of the dipole at its center, as shown in Fig. 10.3.

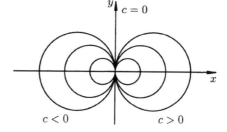

Fig. 10.3. Level lines
of a dipole potential

Problem. Draw the level lines of the potential of a pair of oppositely signed charges of equal magnitude on the plane.

HINT. The difference of the logarithms of two numbers is the logarithm of their ratio. The geometric locus of the points whose distances to two given points are in constant ratio (Fig. 10.4) is a circle (why?). As the charges coalesce ($l \to 0$), the system of circles in Fig. 10.4 becomes the system of circles in Fig. 10.3 with a common tangent at the center of the dipole.

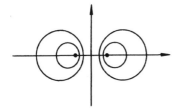

Fig. 10.4. The level lines of the potential of a pair of opposite charges

In our coordinate system a level line of the potential of a dipole, which is a circle, is tangent to the y-axis. If the curve at whose points we are studying the dipole potential is perpendicular to the axis of the dipole and has finite curvature at its center, it has second-order tangency with its osculating circle, which belongs to our family of level lines of the dipole potential. It follows that the limiting value of the dipole potential as a point approaches the center of the dipole along this curve passing through the center is completely determined and equal to the value of the dipole potential on the osculating circle, that is, it equals half the curvature of the curve.

Corollary. *The integrand in the integral that defines a double-layer potential restricted to the torus* $\{x \in S,\ q \in S\}$ *can be extended by continuity even to the diagonal* $x = q$ *(where the integrand is undefined).*

Consider an element of the curve and the angle it subtends at a given point x. The result is a function that is harmonic with respect to x at all points except the points of this element of the curve, as shown in Fig. 10.5.

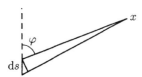

Fig. 10.5. The angle subtended by an element of the curve is a harmonic function

Indeed, this angle is $\dfrac{ds \sin \varphi}{r} = ds\,\dfrac{x}{x^2 + y^2}$, where $x = r \sin \varphi$ and $y = r \cos \varphi$.

From this the *geometric meaning of the (dipole) potential* becomes clear: it is the *element of angle subtended by an element of the curve when seen from a given point.* Moreover, this holds in any dimension: the contribution of the neighborhood ds of the hypersurface to the integral computed at a given point x is the element of solid angle subtended by the element ds at the point x.

Problem. Prove that for any dimension the solid angle subtended by an element of the surface is a harmonic function.

HINT. Use the formula for the derivative of the fundamental solution along a vector field.

For $n = 2$ and $n = 3$ the dipole potential, defined as the derivative of $\ln \frac{1}{r}$ or $\frac{1}{r}$ respectively, coincides exactly with the element of solid angle. In greater dimensions n the coefficient $n - 2$ appears if one uses $u_0 = \frac{1}{r^{n-2}}$. Actually the equality holds only up to a sign depending on the choice of orientation of the hypersurface over which the integration of the angle element is taken. The hypersurface is oriented as the boundary of the oriented "inside" region.

If $\rho \equiv 1$, in the case $n = 2$ we obtain, after integration of the angle element, the "angle subtended by the boundary curve from the point x." More precisely, with a given choice of orientation of the curve the potential is -2π times the number of revolutions of the vector from x to a point on the curve as the entire curve is traversed. When $n = 3$, we obtain similarly the oriented complete solid angle subtended from x by the surface, as shown in Fig. 10.6.

Fig. 10.6. The complete solid angle

This reasoning has replaced some rather lengthy computations. When $n = 2$, we obtain for a double-layer potential of density 1 distributed along a connected closed curve the value -2π in the interior, 0 in the exterior, and $-\pi$ on the curve at each point where a tangent exists.

When $n = 3$, we obtain respectively -4π, 0, and -2π. The values in any dimension can be obtained analogously.

Remark. In the n-dimensional case the level hypersurfaces of the dipole potential are given by the equation $\frac{\sin \varphi}{r^{n-1}} = \text{const.}$

The equipotential surfaces of a dipole have a tangent hyperplane at its center perpendicular to the axis of the dipole, pass through the center of the dipole, and are tangent to one another at this point, but they have a flattening at this point, as shown in Fig. 10.7. When n is even, the equipotential surfaces are analytic, even at the center of the dipole. When n is odd, the equipotential

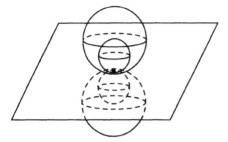

Fig. 10.7. Equipotential surfaces of a dipole when $n > 2$

surfaces have the same kind of singularity at the center of the dipole as the graph of the function $|x|^n$ of $n - 1$ variables.

When $n > 2$, the curvature of every equipotential surface of a dipole equals zero at the center of the dipole.

A surface in general position tangent to the equipotential surfaces at the center of a dipole will therefore have points of intersection with *arbitrarily small* equipotential hypersurfaces of the dipole when $n > 2$. For that reason our dipole potential is very large (in absolute value) on some of them. Hence, on a surface in general position passing through the center of a dipole and perpendicular to its axis the potential of a double-layer element will have a singularity that becomes infinite.

Because of this, in general, the theory of the solvability of boundary-value problems for Laplace's equation in the multidimensional case is not completely analogous to the Fredholm theory normally discussed in textbooks for the case $n = 2$. To be specific, when $n > 2$, the theory of integral equations with continuous kernels used in the Fredholm theory is no longer comprehensive. To be sure, the singularity is a rather weak one; it is integrable, so that one can apply the theory of operators with integrable kernel.

Now suppose the density ρ is not constant. Let us investigate the properties of a double-layer potential.

1. The harmonic property of u in the interior and exterior continues to hold.

2. It is easy to find the difference in the values of u on the boundary of the domain and its limiting values as the boundary is approached from the interior and exterior. It is the same as it would be if the density were constant, namely (when $n = 2$):

$$u_-(x_0) = u(x_0) - \pi\rho(x_0) , \quad u_+(x_0) = u(x_0) + \pi\rho(x_0) ,$$

where $u_-(x_0)$ is the limit from the interior and $u_+(x_0)$ is the limit from the exterior, as shown in Fig. 10.8.

To prove this we consider an auxiliary double-layer potential with constant density $\rho = \text{const} \equiv \rho(x_0)$; the assertion is true for this density. We then consider the potential with density $\rho - \rho(q)$, which equals 0 at q. We shall prove

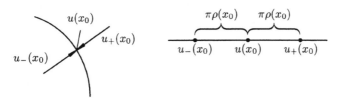

Fig. 10.8. The jump of the double-layer potential

that it has a zero jump. Instead of a formal proof, I shall explain why this is so. If the density were zero in an entire neighborhood of q, that neighborhood would not create a double layer. The rest of the surface would create a double-layer potential in that neighborhood, which is a harmonic function without singularities. Therefore the jumps at q would be zero. In fact the density is not zero in a neighborhood, only at the point q itself. For that reason, in order to prove that the jump vanishes, we need to carry out estimates analogous to those in the appendix to the preceding lecture. These elementary estimates show that the jump in the potential is zero. A sufficient condition for this is that the density be continuous at q.

3. The normal derivative of the double-layer potential (the normal being oriented in the same direction for points in both the interior and the exterior) has no jump.

This is certainly true in the case of a constant potential – the derivative is zero both from within and without. In the general case the derivative of the potential is the force (the "tension" of the field). Consider a cylindrical region Ω near the element ds of the surface, as shown in Fig. 10.9. Let us compute the flux of the force field grad u across the boundary of this region. As in the preceding lecture, we take account of only the normal component:

$$\mathrm{d}s\left(\frac{\partial u}{\partial n_+} - \frac{\partial u}{\partial n_-}\right).$$

Fig. 10.9. Computation of the jump of the normal derivative of a double-layer potential

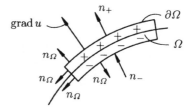

By Stokes' formula, this flux is

$$\int_{\partial\Omega}(\mathrm{grad}\,u, n_\Omega)\,\mathrm{d}s = \int_\Omega \mathrm{div}\,\mathrm{grad}\,u\,\mathrm{d}x = \int_\Omega \Delta u\,\mathrm{d}x = \mathrm{const}\int_\Omega Q\,\mathrm{d}x = 0\,,$$

since this integral is the total charge in the region under consideration, and that is zero (the positive and negative charges cancel each other). Thus

$$\frac{\partial u}{\partial n_+} = \frac{\partial u}{\partial n_-}.$$

We shall use the properties of a double-layer potential below. At the moment we shall study the properties of the Laplacian.

2. Oscillations of spherically symmetric bodies. The fundamental solutions have important applications to the problem of the vibrations of spherically symmetric bodies. Consider, for example, a circle in the plane (a periodic string). The circle $x^2 + y^2 = 1$ is a Riemannian manifold. Hence the Laplacian div grad is defined on it. On the circle there is a standard polar coordinate φ. In terms of this coordinate the operator $\Delta = \text{div grad}$ on the circle can be expressed in the usual way: $\Delta = \dfrac{\partial^2}{\partial\varphi^2}$.

In higher dimensions the problem of a vibrating string is replaced by the problem of a vibrating sphere. There is no single coordinate here. Instead of the trigonometric functions $(\sin k\varphi, \cos k\varphi)$ that describe the natural vibrations of a string, the vibrations of a sphere are described by the so-called *spherical functions*, which arise in all problems of mathematical physics having spherical symmetry. The place of the theory of Fourier series of functions on a circle is taken by expansions in spherical harmonics.

From the practical point of view the problem of the gravitational potential of the Earth, which influences the motion of satellites, is of interest. Satellites, in turn, can be regarded as indicators for judging the distribution of mass in the Earth. As we shall soon see, the contributions of the higher-order harmonics decrease rapidly with distance from the Earth, so that at points not too near the Earth the potential is closely approximated by the sum of a spherically symmetric principal part $\dfrac{c_1}{r}$ and a dipole perturbation $\dfrac{c_2 z}{r^3}$.

Let us consider the problem of the eigenfunctions of the Laplacian on the n-dimensional sphere. For $n = 1$ these are the usual trigonometric functions. By solving the periodic boundary-value problem for the equation $\dfrac{d^2 u}{d\varphi^2} = \lambda u$, one can find all the harmonics. These are just combinations of $\sin k\varphi$ and $\cos k\varphi$ with integers k. (The eigenvalue is $\lambda = -k^2$.)

What is the situation in higher dimensions, for example, on the two-dimensional sphere $x^2 + y^2 + z^2 = 1$? Here we become acquainted with spherical functions, which from this point of view are a generalization of the trigonometric functions.

Consider a function $u(x, y, z)$ that is constant along each ray emanating from the origin. Such a function is said to be homogeneous of degree 0. (We recall that a function is homogeneous of degree k if $u(\lambda x) = \lambda^k u(x)$ for every $\lambda > 0$.)

At the origin the function may be undefined. For example, one homogeneous function of degree 0 is

$$u(x, y, z) = \frac{x^2 + y^2 + z^2}{2x^2 + 3y^2 + 4z^2} \ .$$

Problem. Find $\Delta u\big|_{S^2}$ and compare it with $\operatorname{div}\operatorname{grad}\big(u\big|_{S^2}\big)$.

This last operator is called the *spherical Laplacian*. I shall denote it by $\tilde{\Delta}$.

As it happens, the following identity holds in a space of any dimension n for a homogeneous function u of any degree k:

$$\tilde{\Delta}u = r^2 \, \Delta u - \Lambda u \ , \quad \Lambda = k^2 + k(n-2) \ .$$

The operator on the left-hand side of this equality is the spherical Laplacian, extended to the entire space by homogeneity of degree k. On the unit sphere $r^2 = 1$ it is simply the spherical Laplacian:

$$\tilde{\Delta}u\big|_{r^2=1} = \operatorname{div}\operatorname{grad}\big(u\big|_{r^2=1}\big) \ .$$

Let us now consider some special cases:

1. For $k = 0$ we have the identity

$$\tilde{\Delta}u = r^2 \, \Delta u \ . \tag{10.1}$$

This identity is easily explained. The gradient of a homogeneous function u of degree 0 restricted to the unit sphere coincides with the gradient of the restriction of the function itself to that sphere, since the gradient of a homogeneous function of degree 0 is tangent to the sphere. It is also easy to deduce by considering fluxes that the divergences of the gradients also coincide. That is, for homogeneous functions of degree 0 the usual Laplacian and the spherical Laplacian coincide on the unit sphere: $\big(\tilde{\Delta}u\big)\big|_{r^2=1} = \big(\Delta u\big)\big|_{r^2=1}$. The identity (10.1) can be obtained from the extension of the spherical Laplacian to the entire space preserving homogeneity. Indeed, the operator Δ lowers the degree of homogeneity by 2. It follows that Δu is a homogeneous function of degree -2. In order for the extension to have degree 0, the function Δu of degree -2 must be multiplied by r^2.

2. Let $n = 2$ and $k \neq 0$. Our identity assumes the form

$$\tilde{\Delta}u = r^2 \, \Delta u - k^2 u \ .$$

Let us apply it to the function $u = \operatorname{Re} z^k$, which is harmonic in the plane and homogeneous of degree k. (In particular, when $k = 2$, $u = \cos 2\varphi$ on the unit circle.)

Our identity now assumes the form $\tilde{\Delta}u = -k^2 u$.

For the time being we shall not take up the proof of this general identity. We note only that it is based on Euler's formula for homogeneous functions of degree k (whose proof is left as an exercise):

$$\sum_{i=1}^{n} x_i \frac{\partial u}{\partial x_i} = ku .$$

Corollary. *Suppose a function is homogeneous and harmonic. Then, restricted to the unit sphere, it is an eigenfunction of the spherical Laplacian:*

$$\tilde{\Delta} u = -\Lambda u .$$

We shall see below that all the eigenfunctions of the spherical Laplacian can be obtained in this way.

Definition. The *spherical functions* on the sphere S^{n-1} are the restrictions to the sphere of homogeneous harmonic polynomials in $\mathbb{R}^n$.

Problem. Find the dimension of the space of spherical functions that are the restrictions of the harmonic polynomials of a given degree k of homogeneity in $\mathbb{R}^n$.

For example, for $n = 3$ we obtain:

k	0	1	2	$\cdots$
dimension	1	3	5	$\cdots$
		(basis x, y, z)	(basis xy, yz, zx, $x^2 - y^2, y^2 - z^2$)	

We shall now pause to study an interesting application of spherical functions to a topological problem.

Maxwell's Theorem. *When $n = 3$, all spherical functions of a given degree k can be obtained by successively differentiating the potential $1/r$ along suitable constant vector fields $L_{v_k} \cdots L_{v_1} \dfrac{1}{r}$ and restricting the result to the unit sphere.* (The result of differentiating in this way is called a *multifield potential.*) *Here the k fields $v_1, \ldots, v_k$ are uniquely determined (up to nonzero factors) for a given nonzero spherical function of degree k.*

The dimension of the space of spherical functions of degree k is $2k + 1$. By considering this space up to a nonzero multiplicative constant and omitting 0, we obtain the projective space $\mathbb{R}P^{2k}$.

On the other hand, by Maxwell's theorem all the spherical functions of degree k can be obtained, up to a nonzero multiple, from $1/r$ by differentiating along k constant vector fields of length 1 in $\mathbb{R}^3$. The constant vectors of length 1 are uniquely determined up to sign by the function itself. The differentiations along constant fields commute, so that the result is independent of the order of differentiating.

Thus we have constructed a one-to-one mapping

$$\mathbb{R}P^2 \times \cdots \times \mathbb{R}P^2 / S(k) \to \mathbb{R}P^{2k} \, ,$$

where $S(k)$ is the symmetric group of permutations of k factors.

The first of these spaces is called the *symmetric k^{th} power of the projective plane* $\mathbb{R}P^2$, and is denoted $S^k \mathbb{R}P^2$. The Maxwell mapping $S^k \mathbb{R}P^2 \to \mathbb{R}P^{2k}$ just constructed is a homeomorphism. Maxwell's theorem seems to be the most elementary proof that these spaces are homeomorphic.

A related theorem from algebra is Viète's theorem. The Viète mapping

roots $\longrightarrow$ elementary symmetric polynomials in the roots

defines a homeomorphism of the symmetric k^{th} power of $\mathbb{C}$ and $\mathbb{C}^k$: $S^k\mathbb{C} \approx \mathbb{C}^k$.

The corresponding projective theorem is: $S^k(\mathbb{C}P^1) \approx \mathbb{C}P^k$. The complex projective line $\mathbb{C}P^1$ is the Riemann sphere S^2, so that Viète's projective theorem provides a homeomorphism $S^k(S^2) \approx \mathbb{C}P^k$.

Lecture 11

Spherical Functions. Maxwell's Theorem.
The Removable Singularities Theorem

We consider a homogeneous function F of degree k in $\mathbb{R}^n$:

$$F(\lambda x) = \lambda^k F(x) \quad \forall \lambda > 0 \, .$$

Such a function may be undefined at 0. (A homogeneous function may be defined in a homogeneous domain, for example, in some solid angle with origin at 0.)

We have defined a modified spherical Laplacian $\tilde{\Delta}$ for each k. It maps homogeneous functions of degree k to homogeneous functions of the same degree. We recall its definition: a function is restricted to the unit sphere, then the divergence of its gradient is taken, and the result is extended to all of space except the origin as a homogeneous function of degree k.

Theorem.

$$\tilde{\Delta}F = r^2 \Delta F - \Lambda F \, , \quad \Lambda = k^2 + k(n-2) \, .$$

PROOF. We compute $\Delta F = \operatorname{div} \operatorname{grad} F$. Let f be the restriction of F to the unit sphere: $f = F\big|_{S^{n-1}}$, $f : S^{n-1} \to \mathbb{R}$. Then $(\operatorname{grad} F)\big|_{S^{n-1}} = \operatorname{grad} f + \frac{\partial F}{\partial r}\frac{\partial}{\partial n}$, as shown in Fig. 11.1. (Of course, we are assuming here that the tangent space to the sphere at the point is imbedded in the tangent space to the ambient space.)

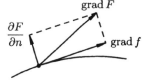

Fig. 11.1. The spherical and total gradients

By homogeneity $F(rq) = r^k f(q)$, so that $\frac{\partial F}{\partial r} = kr^{k-1}f(q)$. It suffices for us to find the gradient on the unit sphere, since it is homogeneous of degree $k-1$.

Our field decomposes into tangential and normal components, the tangential component having no flux across the top or the bottom of a test surface and the normal component having no flux across the lateral walls, as in Fig. 11.2.

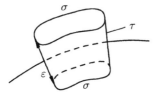

Fig. 11.2. Computation of the flux through a test surface

Let us compute the flux of the gradient across the test surface:

$$\int_{\partial G} (\operatorname{grad} F, \mathbf{n}) \, dS = \int_{\tau} (\operatorname{grad} F, \mathbf{n}) \, dS + \sigma(1 + \varepsilon)^{n-1} k f (1 + \varepsilon)^{k-1} - \sigma k f \,,$$

where the first term is the flux across the lateral surface, the second is the flux across the top, and the third is the flux across the bottom. As a result, we obtain:

$$\int_{\tau} (\operatorname{grad} F, \mathbf{n}) \, dS + \sigma(n + k - 2)\varepsilon k F + o(\varepsilon) \,.$$

Here $\varepsilon\sigma$ is the volume of the test region. The functions F and f are equal on the unit sphere. Dividing the flux by the volume of the test region and contracting the test region to a point on the unit sphere, we find that at this point

$$\Delta F = \tilde{\Delta} f + k F(n + k - 2) \,.$$

Indeed, the flux of $\operatorname{grad} F$ across ∂G is the integral of $\operatorname{div} \operatorname{grad} F$ with respect to G, while the flux of $\operatorname{grad} F$ across τ is $\varepsilon \times$ (the flux of $\operatorname{grad} f$ across $\partial \sigma$)$+o(\varepsilon)$. This last flux is the integral of $\operatorname{div} \operatorname{grad} f$ over σ.

Thus, our identity is proved on the unit sphere. When it is extended to a point at distance r from the origin, the F on the right-hand side is multiplied by r^k, while ΔF is multiplied by r^{k-2}. That is why the factor r^2 arises in the first term of the identity being proved.

The modified spherical Laplacian is homogeneous of degree k by definition. Therefore if the identity holds on the unit sphere, it holds everywhere (except at the origin). Thus the identity is proved. □

Our theorem is a special case of the following simple, but useful remark on the Laplacian of a function on a submanifold of Euclidean space. In computing the value of the Laplacian at a point one can replace the submanifold by its tangent space at that point and the function by the corresponding function in the tangent space.

Consider an m-dimensional submanifold ("surface")

$$y = f(x) , \quad x \in \mathbb{R}^m , \quad y \in \mathbb{R}^l$$

in the Euclidean space $\mathbb{R}^{m+l}$ with the metric $\mathrm{d}x^2 + \mathrm{d}y^2$. We call the x-subspace *horizontal* and the y-subspace *vertical*. We shall take the Cartesian coordinates of x as the coordinates of the point $X = (x, y)$, $y = f(x)$, on the surface.

A function U defined on the surface can be written in these coordinates as a function on the horizontal plane

$$u(x) = U(X) .$$

Lemma. *Assume that the tangent plane to the surface at the point 0 is horizontal (that is, that $(\mathrm{d}f/\mathrm{d}x)(0) = 0$). Then the Riemannian Laplacian of the function U at this point on the surface coincides with the Euclidean Laplacian of the function u on the tangent plane.*

PROOF. We write the metric on the surface as a Riemannian metric $\mathrm{d}S^2$ on the horizontal plane. It follows from the Pythagorean theorem that the difference $\mathrm{d}y^2$ between this metric and the Euclidean metric in the horizontal plane is a second-order infinitesimal in $|x|$:

$$\mathrm{d}S^2 - \mathrm{d}x^2 = O(|x|^2) .$$

(Here and below smallness of a quadratic form or differential operator means, of course, that the coefficients are small.) Therefore the operators GRAD and DIV of the Riemannian gradient and the Riemannian divergence corresponding to the metric $\mathrm{d}S^2$ on the horizontal plane differ from the Euclidean grad and div by infinitesimals of second order:

$$\mathrm{GRAD} - \mathrm{grad} = O(|x|^2) , \quad \mathrm{DIV} - \mathrm{div} = O(|x^2|) .$$

Applying these operators successively to the function u, we find

$$\mathrm{DIV\,GRAD}\,u = \mathrm{div\,grad}\,u + O(|x|^2) + \mathrm{div}\,O(|x|^2) .$$

This last term is $O(|x|)$, so that

$$(\mathrm{DIV\,GRAD}\,u)(0) = (\mathrm{div\,grad}\,u)(0) . \qquad \square$$

Remark. The translation of a function to the tangent plane does not necessarily have to be done using orthogonal projection: one can use any family of smooth curves. All that matters is that the curve passing through the point is orthogonal to the surface at that point.

Example. Consider the unit sphere $y = \sqrt{1 - x^2}$ in n-dimensional Euclidean space and the point $x = 0$ on it ($m = n - 1$, $l = 1$). Let us consider the function U on the sphere and compute its spherical Laplacian at $x = 0$.

By the lemma it equals the Euclidean Laplacian $\Delta(x)u$ (the sum of the second partial derivatives of the corresponding function $u(x)$ with respect to x_i).

We shall denote the homogeneous extension of U of degree k to Euclidean space by $\widetilde{U}$. On the tangent plane $y = 1$ to the sphere at the point $x = 0$ this function becomes

$$\tilde{u}(x) = \left(\sqrt{1 + x^2}\right)^k u(\tilde{x}) , \quad |\tilde{x} - x| = O(|x|^3) .$$

Therefore the Euclidean Laplacian of the extension at the point $x = 0$, $y = 1$ is

$$\Delta\widetilde{U} = \frac{d^2}{dy^2} y^k u(0) + (\Delta_x \tilde{u})(0) = k(k - 1)u(0) + \Delta_x \tilde{u} .$$

But $\tilde{u}(x) = u(x) + \dfrac{k}{2} x^2 u(0) + O(|x|^3)$. Therefore at the point $x = 0$

$$\Delta_x(\tilde{u}) + \Delta_x u + k(n - 1)u .$$

Finally, at the point in question,

$$\Delta\widetilde{U} = \Delta_x u + [k(n - 1) + k(k - 1)]u(0) .$$

Since the point chosen differs in no way from the others, we have proved our formula $\widetilde{\Delta}F = r^2 \Delta F - \Lambda F$, $\Lambda = k(n + k - 2)$ for $r = 1$. By homogeneity it holds everywhere.

Corollary. *If the function F (defined in $\mathbb{R}^n \setminus \{0\}$) is harmonic and homogeneous of degree k, it is an eigenfunction for the (modified) spherical Laplacian: $\widetilde{\Delta}F = -\Lambda F$ (that is, its restriction to the sphere is an eigenfunction of the spherical Laplacian). Conversely, an eigenfunction of the spherical Laplacian, when extended from the unit sphere as a homogeneous function of degree k, is harmonic everywhere except at the origin.*

We remark that for a given eigenfunction of the Laplacian on the unit sphere one can find *two* harmonic extensions, since the quadratic equation $\Lambda = k^2 + k(n - 2)$ determines *two* values of k.

Example. Let $n = 2$, that is, we are considering functions on the plane and their restriction to the circle. We have $\Lambda = k^2$. The function x is a harmonic homogeneous function of degree 1. But there also exists a harmonic homogeneous function of degree -1, whose restriction to the unit circle is the same as that of the function x.

In polar coordinates the formulas of these conjugate functions: $F = r\cos\varphi = x$, $\hat{F} = (\cos\varphi)/r = x/r^2$, form a dipole potential (the kernel of a double-layer potential). Similarly, for the conjugate exponents of homogeneity k and $-k$:

$$F = r^k \cos k\varphi , \quad \hat{F} = \frac{\cos k\varphi}{r^k} .$$

Let $n = 3$, that is, we are considering functions of three variables and their restrictions to the two-dimensional sphere. We have $\Lambda = k^2 + k$. Therefore the exponent conjugate to k is $\hat{k} = -1 - k$.

For $k = 0$ the conjugate functions are $F = 1$, $\hat{F} = 1/r$.

For $k = 1$ we have $F = z$, $\hat{F} = z/r^3$. As in the two-dimensional case, this is a dipole potential (the kernel of a double-layer potential).

For $k = 2$ we can find a harmonic quadratic form, for example, $F = \frac{1}{2}(3z^2 - x^2 - y^2 - z^2)$, $F|_{S^2} = f = \frac{1}{2}(3z^2 - 1)$.

The functions z and f are spherical, that is, they describe the natural vibrations of a sphere. Moreover, they are so-called *zonal* functions: they are invariant under rotation about the z-axis and reverse sign when crossing the parallels that divide the sphere into separately vibrating zones, and the dividing parallels remain fixed, as shown in Fig. 11.3.

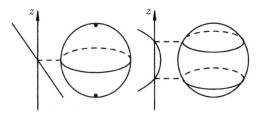

Fig. 11.3. Zonal functions describe the natural vibrations of a sphere

Problem. Does there exist, for every integer k, a zonal spherical function that is the restriction to the sphere of a polynomial of degree k in z?

In the general case $\Lambda = k^2 + k(n - 2)$, and the homogeneity exponent conjugate to k is $\hat{k} = 2 - n - k$. The values of the conjugate functions at the points rq (q being a point of the unit sphere and r the distance from the origin) are $F(rq) = r^k f(q)$ and $\hat{F}(rq) = f(q)/r^{k+n-2}$.

These values can be connected using an inversion. When this is done, the image of qr is q/r, as shown in Fig. 11.4.

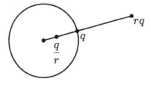

Fig. 11.4. Inverse points

We have: $F(q/r) = f(q)/r^k$, so that $\hat{F}(rq) = F(q/r)/r^{n-2}$. (We remark that when $n = 2$ the inversion continues to be harmonic, and the coefficient drops out of the last formula.)

For any n the exponent of homogeneity k has completely disappeared from the formula connecting $F(q/r)$ and $\hat{F}(rq)$. Hence this formula holds for linear combinations of homogeneous functions with different exponents of homogeneity. But any harmonic function can be approximated with such functions. Thus under this transformation any harmonic function F maps to a harmonic function $\hat{F}(x) = F(x/|x|^2)/|x|^{n-2}$. (This can also be verified by direct computation.)

Proposition. *There exist homogeneous harmonic functions of degree k only for integers k. (Eigenfunctions of the spherical Laplacian exist only for Λ corresponding to integer k.)*

PROOF. We begin by studying homogeneous harmonic polynomials. The space of homogeneous polynomials of degree zero is one-dimensional, those of degree 1 have dimension n. We shall find the dimension of the space of polynomials of degree 2 and higher in the three-dimensional case for the sake of simplicity (see Fig. 11.5):

$$
\begin{array}{lcccc}
\text{degree} & 0 & 1 & 2 & 3 \\
\text{dimension} & 1 & 3 & 6 & 10
\end{array}
$$

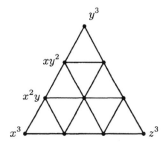

Fig. 11.5. Determination of the dimension of the space of homogeneous polynomials of degree three

In general $\dim S^k \mathbb{R}^3 = C^2_{k+2} = \dfrac{(k+2)(k+1)}{2}$. The Laplacian maps this space to the space $S^{k-2}\mathbb{R}^3$ of homogeneous polynomials of degree 2 less than the degree of the polynomials of the original space. The harmonic homogeneous polynomials of degree k form a vector space, namely the kernel of the mapping

$$\Delta : S^k \mathbb{R}^3 \to S^{k-2}\mathbb{R}^3 .$$

Claim. *The Laplacian maps $S^k \mathbb{R}^3$ onto $S^{k-2}\mathbb{R}^3$.*

Indeed, any polynomial is the image of some polynomial under the Laplacian. (It suffices to verify this for monomials. For monomials in one variable x this is obvious: $\partial^2 u/\partial x^2 = x^p$ when $u = x^{p+2}/(p+2)(p+1)$. For the monomial $x^p y^q$ in x and m additional variables y_i we begin with the monomial

$u = x^{p+2}y^q/(p+2)(p+1)$. We have $\Delta_{x,y}u = x^p y^q + v$, where $v = \Delta_y u$ is a polynomial whose degree in y is less than $|q|$. Therefore $x^p y^q$ will be in the image of $\Delta_{x,y}$ if all polynomials of degree less than $|q|$ in y belong to that image. This makes it possible to prove what is required by induction on $|q|$. (If $\Delta_{x,y}w = v$, then $\Delta_{x,y}(u - w) = x^p y^q$.)

Thus the dimension of the space of spherical functions of degree k equals the difference of the dimensions of $S^k \mathbb{R}^3$ and $S^{k-2}\mathbb{R}^3$, that is,

$$\frac{(k+2)(k+1)}{2} - \frac{k(k-1)}{2} = 2k + 1 . \qquad \square$$

Corollary. *For each nonnegative integer k there exists a $(2k+1)$-dimensional vector space of spherical functions on the sphere S^2. These are eigenfunctions of the spherical Laplacian with eigenvalue $-\Lambda$, where $\Lambda = k^2 + k$, and they are the restrictions to the sphere of harmonic homogeneous polynomials of degree k in $\mathbb{R}^3$.*

PROOF. The fact that the restrictions are eigenfunctions follows from the identity just proved. The dimension of the space of harmonic homogeneous polynomials of degree k in $\mathbb{R}^3$ has just been computed. The dimension of the space of their restrictions to the sphere is the same, since a homogeneous polynomial that vanishes on the sphere is identically zero. $\square$

We have now found the dimension of the space of harmonic homogeneous polynomials of degree k for $n = 3$. In the general case the dimension grows like k^{n-2}.

Claim. *Spherical functions with different eigenvalues are pairwise orthogonal.*

Indeed the natural vibrations with different natural frequencies are always orthogonal, like the principal axes of an ellipsoid with different lengths in Euclidean space.

Hence it follows, for example, that $\int_{S^2} (3z^2 - 1)\,\mathrm{d}z = 0$. Of course, this integral can be computed explicitly using symmetrization: the three integrals for x, y, and z are all equal, and their sum equals 0.

Theorem. *Every spherical function (eigenfunction of the spherical Laplacian) is the restriction to the sphere of a homogeneous harmonic polynomial in the ambient space.*

PROOF. The eigenvalues of the spherical Laplacian are nonpositive, since the potential energy – the Dirichlet integral – is nonnegative.

Eigenfunctions that are harmonic on the sphere correspond to the eigenvalue 0. A function that is harmonic on the sphere is constant (by the maximum principle), since the sphere is a closed connected manifold (compact and without boundary). Indeed, let us cut a small hole out of the sphere. Then the maximum will be reached on the boundary of the hole. Contracting the hole to a point, we verify that at every point on the manifold the value of a function harmonic on a closed connected manifold equals the maximum of the function on the whole manifold.

Now let us consider any negative eigenvalue $-\Lambda$, where $\Lambda > 0$. For any $\Lambda > 0$ there exists an exponent of homogeneity $k > 0$ such that $\Lambda = k(k + n - 2)$, as shown in Fig. 11.6.

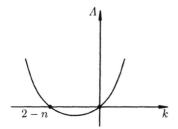

Fig. 11.6. The eigenvalue as a function of the degree

An eigenfunction of the spherical Laplacian with eigenvalue $-\Lambda$ can be extended to the entire space with 0 removed as a harmonic homogeneous function of degree k (by our fundamental identity).

Since k is positive, the extended function is bounded in a neighborhood of 0. By the removable singularities theorem to be proved below, a bounded harmonic function in a neighborhood of 0 remains harmonic when it is extended to the origin by continuity. For the extended function we have $u(0) = 0$, since $u(rq) = r^k u(q)$, $k > 0$.

The extended function is harmonic at 0 and consequently infinitely differentiable. But a homogeneous function of degree k that is infinitely differentiable at 0 is necessarily a homogeneous polynomial of degree k (and k is necessarily a nonnegative integer). This can be seen by Taylor's formula. Thus, any spherical function can be extended to a homogeneous harmonic polynomial in the entire space.   □

Maxwell's Theorem. *When $n = 3$, all spherical functions of degree k can be obtained by successive differentiation of the function $1/r$ along k directions.*

PROOF. First of all, it is obvious that the derivative of a harmonic function along a constant vector field is harmonic. For example, the partial derivatives of a harmonic function are harmonic.

Second, on the sphere the derivatives coincide with certain homogeneous polynomials. This is true for the original function $1/r$: that is, 1 is a homogeneous polynomial of degree 0.

Suppose that after a differentiations the result is a function of the form F/r^a, where F is a homogeneous polynomial of degree d. When we differentiate along a constant vector field the homogeneity is preserved, while the degree decreases by 1. Let $\nabla_{\mathbf{v}}$ be differentiation along the constant field $\mathbf{v}$. Then

$$\nabla_{\mathbf{v}}\frac{F}{r^a} = \frac{\nabla_{\mathbf{v}}F\cdot r^a - Fa\cdot r^{a-2}(\mathbf{v},\mathbf{r})}{r^{2a}} = \frac{(\nabla_{\mathbf{v}}F)r^2 - aF(\mathbf{v},\mathbf{r})}{r^{a+2}}.$$

The numerator here is a homogeneous polynomial of degree $d+1$. Therefore the resulting harmonic function, after being restricted to the unit sphere, coincides with a homogeneous polynomial. This polynomial is harmonic. Indeed, after differentiating we obtain a harmonic homogeneous function of negative decree. Its exponent of homogeneity is conjugate to the exponent of homogeneity of the numerator, which is a homogeneous polynomial of positive degree. When a harmonic function is restricted to the unit sphere, the result is a spherical function. If we extend it with the conjugate positive exponent of homogeneity, the result is a polynomial and harmonic in the entire space (by the preceding theorem). Thus, the restriction of the derivative in question to the sphere coincides with the restriction of a homogeneous polynomial that is harmonic in the entire space.

It can be proved (see Appendix A) that the space of functions produced in this way is a vector space (which is not at all obvious).

To prove that all spherical functions can be obtained in this way it suffices to show that no eigenspace of the space of spherical functions can be invariant with respect to all rotations. Then the subspace obtained by the Maxwell construction (which is invariant under rotations) must coincide with the entire space of spherical functions.

To prove that the representation of the group of rotations of the sphere S^2 by linear transformations of the $(2k+1)$-dimensional space of spherical functions is irreducible, that is, that there are no nontrivial invariant subspaces, it suffices, for example, to verify that the entire space consists of linear combinations of a single function and its rotations. In the role of such a spherical function that generates the whole space we can take, for example, a zonal spherical function. Let us study these functions in somewhat more detail.

There are two ways of extending spherical functions to homogeneous harmonic functions. Let us consider the case of the Maxwell construction, in which all differentiations are carried out with respect to the same direction, for example, along the z-axis. Restricting the function $\left(\frac{\partial}{\partial z}\right)^k\frac{1}{r}$ to the sphere, we obtain the spherical functions. When $k = 0$, we have 1; when $k = 1$ we obtain z/r^3; each time we shall obtain expressions of the form

$$\left(\left(\frac{\partial}{\partial z}F\right)r^2 - aFz\right)\Big/r^{a+2}.$$

If $F = F(z,r)$, this form of F is preserved under these differentiations, so that a function of z alone results on the unit sphere. Hence, *in the space of spherical functions of any degree k there is a function depending on z alone*. It is the restriction to the sphere of a certain polynomial in z, which is called a *Legendre polynomial*. Legendre polynomials of different degrees are pairwise orthogonal on the interval $[-1, 1]$ of the z-axis, since the integral over the sphere of their product, which equals zero, transforms easily into an integral with respect to z, by Archimedes' theorem that a sphere without poles and a cylinder are symplectomorphic. (The element of area of the sphere equals the element of area of the circumscribed cylinder under projection onto the cylinder along horizontal radii, as shown in Fig. 11.7.) □

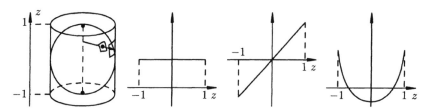

Fig. 11.7. Archimedes' symplectomorphism of a sphere onto a cylinder and the graphs of the first three Legendre polynomials

Corollary. *The Legendre polynomial of degree k has k distinct roots in the interval $(-1, 1)$.*

Indeed, the number of zeros is at most k, since the polynomial has degree k. Assume that the number of roots is less than k. Then the number m of roots on $(-1, 1)$ at which the polynomial changes sign is also less than k. We form a linear combination of Legendre polynomials of degree at most m having a zero of order one at these m zeros. There exists such a linear combination, since the linear combinations of Legendre polynomials of degree at most m constitute *all* polynomials of degree at most m. (The space of polynomials has degree $m + 1$, and the first $m + 1$ Legendre polynomials are linearly independent.) We multiply this linear combination by our Legendre polynomial of degree k. The product does not change sign on the interval $(-1, 1)$, which contradicts the orthogonality of the Legendre polynomial of degree k to all Legendre polynomials of lower degree.

Thus the number of distinct roots of the Legendre polynomial of degree k on the interval $(-1, 1)$ is k. Consequently these roots are simple roots.

Every Legendre polynomial describes a zonal natural vibration of the sphere, as shown in Fig. 11.8.

Besides the Legendre polynomial f of degree k, same eigenvalue also has the associated functions $f_j(z) \cos j\varphi$, $f_j(z) \sin j\varphi$ $(j < k)$, which also describe natural vibrations, as in Fig. 11.9; each piece of the grid vibrates independently.

The associated functions can be obtained from the Legendre polynomial by rotating the sphere about the x-axis or the y-axis through a small

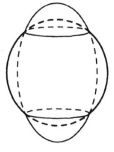

Fig. 11.8. The vibrations of a sphere corresponding to a Legendre polynomial

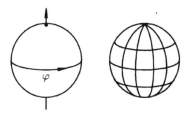

Fig. 11.9. The vibrations of a sphere corresponding to the associated functions

angle α. The Legendre polynomial thereby becomes a nearby spherical function. An associated function measures the difference between the original function and the translated function. To be specific, the derivative of the translated function with respect to α is an associated spherical function: $\bigl(x(\partial/\partial z) - z(\partial/\partial x)\bigr)f(z) = xf'(z) = f'(z)\sqrt{1 - z^2}\cos\varphi$. Repeated differentiation produces $\cos 2\varphi$ and so forth. These computations prove, by the way, that all spherical functions are generated by a single zonal function and consequently prove that the representation is irreducible (has no invariant subspaces).

In particular, it follows from this that the Maxwell construction produces all spherical functions.

It can be proved (see Appendix A below) that the space of kth derivatives of the function $1/r$ with respect to k constant (and consequently commuting) vector fields is a vector space. (The derivative is a linear function of the field.) The dimension of this vector space of derivatives is at most $2k + 1$, since a nonzero k^{th} derivative is determined by the directions of k vectors (k points on the 2-sphere) and one common coefficient.

On the other hand, we constructed above a linear transformation from this vector space of dimension $2k + 1$ onto the whole $(2k + 1)$-dimensional space of harmonic homogeneous polynomials of degree k in $\mathbb{R}^3$. Thus this transformation is an isomorphism. Consequently the Maxwell construction not only gives all spherical functions with a given k, it gives them each exactly once, up to a multiple. This proves the following topological proposition:

$$S^k(\mathbb{R}P^2) \approx \mathbb{R}P^{2k} .$$

A theory of spherical functions exists in any dimension, and there are zonal functions in every dimension. When $n = 3$, these are the Legendre polynomials. Let us see what results when $n = 2$.

The vibrations of a circle are described by the eigenfunctions

$$\cos k\varphi , \quad \sin k\varphi .$$

The existence of zonal vibrations means that a linear combination of these two functions can be formed that depends only on z. But that is obvious: such a function is the function $z = \cos\varphi$, and also the function $\cos k\varphi$ for any k, as in Fig. 11.10.

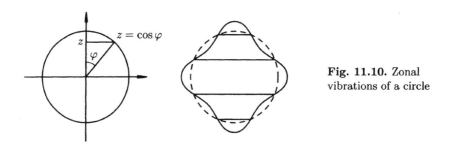

Fig. 11.10. Zonal vibrations of a circle

Corollary. $\cos k\varphi = T_k(\cos\varphi)$, *where T_k is a polynomial.*

These polynomials are called the *Chebyshev polynomials*. They are orthogonal on the interval $[-1, 1]$, with a weight, however (compute which weight) and have zeros with properties analogous to those of the Legendre polynomials.

Incidentally, the role of the associated functions here is played by the functions $\sin k\varphi$ resulting from zonal functions by differentiation with respect to the angle of revolution of the zonal function.

The majority of the so-called special functions of mathematical physics arises in the problem of the vibrations of bodies with various symmetries in a space of some dimension.

For example, the problem of the vibrations of a disk lead to the so-called Bessel functions. Bessel was a German astronomer who encountered these functions in Fourier-series expansions of the gravitational potential of the Sun restricted to a Keplerian ellipse.

The fact that such important physical problems as the problem of vibrations of a disk and the attraction of planets lead to the same mathematical theory is a marvelous manifestation of the universality of mathematics and the unity of all reality.

To conclude the proof Maxwell's striking theorem, it remains only to prove the (also striking) removable singularities theorem used in it.

Theorem. *Let $n = 2$. If a function in $\mathbb{R}^n$ is harmonic in some punctured neighborhood of a point and bounded, it can be extended by continuity to that point and the extended function is harmonic in the entire neighborhood of the point, including the point itself.*

This theorem has a simple physical meaning. A harmonic function describes the stationary state of a membrane stretched over some curve. For example, the membrane could be stretched onto two hoops. The theorem asserts that the membrane cannot be propped up at one point by a needle. The needle would pass right through, as in Fig. 11.11.

Fig. 11.11. A membrane can be propped up by a hoop, but not by a needle

In the case of a thin plate, which is described by the equation $\Delta^2 u = 0$, the corresponding assertion is not true, that is, a plate can be supported at one point.

PROOF. Consider a circle with center at the point 0 in question. (By translating and dilating coordinates we can assume that the circle is defined by the condition $r = 1$.) We construct a harmonic function in the disk that coincides with the given function on the boundary. We already know that such a function exists. (An explicit function is given by the Poisson integral, exhibited in Lecture 7.) We must prove that the difference between the original function and the constructed function, which we denote u, is zero in the punctured disk.

Consider the function $u_0 = C \ln(1/r)$. It is harmonic in the punctured disk and equal to zero on the circle $r = 1$. We choose C so that this function is greater than u when $r = \varepsilon$.

The function u is bounded in the punctured disk; suppose $|u| \leq M$ there. So that the function u_0 will be equal to M when $r = \varepsilon$ we choose $C = M/\ln(1/\varepsilon)$. Then $u = u_0 = 0$ on the unit circle and $u \leq u_0$ on the circle of radius ε. Both functions are harmonic in the annulus between the circles and $u \leq u_0$ on both boundaries. By the maximum principle the function $u_0 - u$ is nonnegative everywhere in the annulus, as shown in Fig. 11.12. Thus

$$0 \leq u \leq \frac{M \ln(1/r)}{\ln(1/\varepsilon)} .$$

The function $-u$ also satisfies such an inequality.

Let ε tend to 0. Then we obtain $u = 0$ at all points of the disk, as in Fig. 11.12. $\square$

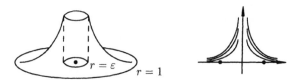

Fig. 11.12. The barrier method of removing a singularity

An analogous proof carries over to the case when the singularity at zero is weaker than logarithmic. The method just described for removing the singularity is called the *barrier method*. The essence of the method is to find a positive harmonic function away from a singularity tending to infinity as the singularity is approached. By multiplying such a "barrier" function by a suitable constant, we can make it as small as desired at any point away from the singularity. If the function being studied tends to infinity more slowly than the barrier function as the singularity is approached, the singularity is removable: on the boundary of an annulus the function in question is less (not greater) than the barrier function, so that it is less (not greater) than the barrier function everywhere and hence equal to zero.

The removable singularities theorem holds in any dimension; the singularity must be weaker than that of the fundamental solution. The proof is exactly the same as in the case $n = 2$, except that $C \ln(1/r)$ must be replaced by $C(r^{n-2} - 1)$.

Corollary. *A bounded harmonic function in the plane is constant.*

Remark. This result, due to Liouville, also holds in $\mathbb{R}^n$ for any n, but the simple proof given below is valid only for $n = 2$.

PROOF. By use of stereographic projection the Riemann sphere can be covered by two flat charts related to each other by the transformation $w = 1/z$. Consider the function $u(z)$ in the chart w, where it is harmonic and bounded in a neighborhood of zero. By the removable singularities theorem it is bounded and harmonic on the entire Riemann sphere and hence constant.   □

Remark. It suffices to have boundedness from only one side. For example, every positive harmonic function on the whole plane is constant.

Problem. If a harmonic function on the plane grows no faster than a polynomial, then it is a polynomial.

HINT. The derivatives of such a function also grow no faster than polynomials. Hence, by the Cauchy–Riemann equations it follows that the holomorphic function f whose real part is the original harmonic function grows no faster than some polynomial. Therefore the function $f(z)/z^N = w^N f(1/w)$, which is holomorphic in some punctured neighborhood of $z = \infty$, (that is, $w = 0$) is bounded. It follows from the removable singularities theorem that it is holomorphic at $w = 0$, so that f is a polynomial of degree at most N.

Problem. Does a function that is harmonic and bounded in the complement of a line segment in $\mathbb{R}^3$ have a removable singularity?

HINT. Take the potential of a charge of density 1 on the line segment as the barrier function.

Problem. Prove Liouville's theorem in $\mathbb{R}^n$: a bounded harmonic function is constant.

HINT. Use Poisson's formula or an expansion in spherical functions on a large sphere.

Lecture 12

Boundary-Value Problems for Laplace's Equation. Theory of Linear Equations and Systems

12.1. Four Boundary-Value Problems for Laplace's Equation

Consider the compact connected smooth hypersurface S^{n-1} in $\mathbb{R}^n$, which divides $\mathbb{R}^n$ into two regions: the interior (bounded) region G and the exterior (unbounded) region G'. Suppose a continuous function $f : S^{n-1} \to \mathbb{R}$ is given on the boundary. The Dirichlet problem for Laplace's equation is to find a function u in the closure of the region G (G') for which the following conditions hold.

1. The function u is harmonic: $\Delta u = 0$ in the region G (the interior problem) or G' (the exterior problem).
2. The function is continuous in the closure of the region: $u \in C(\overline{G})$ (resp. $u \in C(\overline{G'})$).
3. The function satisfies the boundary condition $u\big|_{S^{n-1}} = f$.
4. In the case of the exterior problem certain additional conditions are considered at infinity; these conditions affect the existence and uniqueness of the solution. They are differently posed in different textbooks. The most common condition (which looks strange at first sight) is

$$|u| < C \quad \text{as } x \to \infty, \ n = 2 ;$$
$$u \to 0 \quad \text{as } x \to \infty, \ n > 2 .$$

The *Neumann problem* is this problem with Condition 3 replaced by $\frac{\partial u}{\partial n} = f$; in that case in Condition 2 it is necessary to require C^1-smoothness or some other sufficient condition for the normal derivative to exist.

In the case of the Neumann problem conditions like Condition 4 are usually imposed as well. Physically the Dirichlet boundary condition (Condition 3) corresponds to a membrane clamped along its rim and the Neumann condition $\frac{\partial u}{\partial n} = 0$ to a free membrane.

By combining the interior and exterior versions of the Dirichlet and Neumann problems we obtain four types of boundary-value problems. Let us take each of them up in turn.

1. The Interior Dirichlet Problem

There exists a unique solution to this problem.

For example, when $n = 2$, the Poisson integral gives the solution of the Dirichlet problem for the disk. By the Riemann mapping theorem any region can be conformally mapped onto the disk, so that it is possible to obtain a solution of the Dirichlet problem there also. Unfortunately, the proof of the Riemann mapping theorem is based on the existence of a solution of the Dirichlet problem; and besides, this technique does not apply in higher dimensions. Nevertheless, the result holds in any dimension: There exists a unique solution of the Dirichlet problem.

This result remains valid in the case of a planar or multidimensional region bounded by several connected surfaces. The uniqueness follows immediately from the maximum principle (see the preceding lectures). The idea of the proof of existence is to minimize the Dirichlet integral under prescribed boundary conditions. This minimum can be achieved, as we already know, only at a solution of the Dirichlet problem. It is in fact attained, as is physically obvious. (It is the stationary state of a membrane stretched over the boundary.) But the proof is nontrivial and subtleties are encountered. For example, for a membrane propped up at one point the lower bound of the Dirichlet integral is not attained (see the removable singularities theorem in Lecture 10). We shall not prove that the minimum exists.

2. The Exterior Dirichlet Problem

This problem can be reduced to the interior problem. To do so one must perform an inversion with center at some point of the region. If this point is taken as the origin, the inversion is given by the formula $x \mapsto x/|x|^2$. When $n = 2$, harmonic functions transform to harmonic functions. For $n > 2$ the function $u(x/|x|^2)/x^{n-2}$ will be harmonic. We study the new function in the bounded region into which the exterior of G maps under the inversion, as shown in Fig. 12.1, and pose the interior Dirichlet problem for this function.

One can see that a singularity may arise at zero. But if the original function is bounded outside of G, the singularity is removable. Thus there exists a unique bounded solution of the Dirichlet problem. When $n > 2$, this solution tends to zero at infinity, since the bounded solution is divided by the quantity $|x|^{n-2}$ when it is returned to the exterior region, and that quantity grows at infinity.

Fig. 12.1. Transformation of a region under inversion

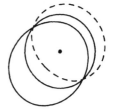

An unbounded solution is not unique. For example, when $n = 2$, if the boundary is the circle $r = 1$, one can add any multiple of the fundamental solution (which grows at infinity and equals zero on the boundary) to the bounded solution. One can also add other harmonic functions, for example $r \cos \varphi$, $(\cos \varphi)/r$ and the like. If we omit Condition 4, we obtain an infinite-dimensional space of solutions.

When $n > 2$, Condition 4 is chosen so that the transformed function $u(x/|x|^2)/x^{n-2}$, when divided by the fundamental solution $1/|x|^{n-2}$ tends to zero as $|x| \to 0$. The removable singularities theorem is then applicable to the transformed function, and the exterior Dirichlet problem reduces to the interior problem. That is how the condition $u \to 0$ as $x \to 0$ is obtained.

3. The Interior Neumann Problem

Suppose first that $n = 2$. A harmonic function is the real part of a holomorphic function. The imaginary part of this holomorphic function is the harmonic conjugate of the real part. The Cauchy–Riemann equations imply that the derivative of the unknown function along the normal to the boundary equals the derivative of the conjugate function in the orthogonal direction, as shown in Fig. 12.2.

Fig. 12.2. Reduction of the Neumann problem to the Dirichlet problem

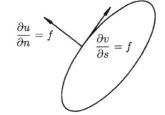

We thus find that the derivative of the conjugate function v in the direction tangential to the boundary is prescribed. If v itself could be recovered from this derivative along the boundary, the Neumann problem for u would have been reduced to the Dirichlet problem for v. But it is not always possible to recover v from its derivative along the boundary, since topological obstacles may intervene. For example, if the boundary is topologically a circle with coordinate s, a necessary and sufficient condition for v to be recoverable from $f = \partial v/\partial s$ is that $\oint_{\partial G} f \, ds = 0$. In the case of a nonsimply connected region, an annulus, for example, it remains true that only one condition is required for solvability: the integral of the boundary function over the boundary of the annulus must be equal to zero. But there are not two conditions asserting that the integral over each circle must vanish, as one might have expected. (It is instructive to think about why this is so!) Thus, the following result is obtained.

Theorem. *The interior Neumann problem is solvable if and only if $\oint_{\partial G} f \, ds = 0$. The solution is unique up to an additive constant. Thus the solution space is one-dimensional.*

Remark. Consider a linear transformation $A : X \to Y$. In the space X there is a subspace $\operatorname{Ker} A$, the kernel of A, and in Y there is a subspace $\operatorname{Im} A$, the image of A.

The number $k = \dim \operatorname{Ker} A$ is the number of linearly independent solutions of the homogeneous equation $Au = 0$, and the number $s = \operatorname{codim} \operatorname{Im} A$ is the number of independent conditions one must impose on the right-hand side of the nonhomogeneous equation $Au = f$ to guarantee solvability.

We also use the notation $c = \operatorname{codim} \operatorname{Ker} A$ and $i = \dim \operatorname{Im} A$. We have $c + k = \dim X$ and $i + s = \dim Y$. In addition the equality $c = i$ holds. (The quotient space over the kernel is isomorphic to the image.) Thus $k - s = \dim X - \dim Y$.

We see that the number $k - s$, called the *index* of A is actually independent of the transformation and is determined by the spaces alone.

Sometimes this observation can be generalized to infinite-dimensional spaces. This can be done provided A is sufficiently well approximated by finite-dimensional transformations. In our applications X and Y will be function spaces, and we shall require sufficiently good approximation by finite sums of Fourier series.

Let us examine our boundary-value problems from this point of view.

For the interior Dirichlet problem $s = 0$, $k = 0$, $s - k = 0$.

For the interior Neumann problem $s = 1$, $k = 1$, $s - k = 0$.

In both cases, it turns out to be a question of mapping the same spaces.

The necessity of the condition $\oint_{\partial G} f \, ds = 0$ for solvability of the interior Neumann problem can also be obtained starting from considerations of minimizing the Dirichlet integral. For $n = 2$ it is obvious, for example, that the normal derivative cannot be positive on the entire boundary without violating the maximum principle, as shown in Fig. 12.3. If the derivative is positive, a singularity that is at least logarithmic arises, as shown in Fig. 12.3.

Fig. 12.3. The normal derivative of a harmonic function cannot be positive

Problem. Integrate by parts in the Dirichlet integral $\int_G (\nabla u)^2 \, dx$ in the multidimensional case and verify that the condition $\oint_{\partial G} f \, ds = 0$ is necessary for solvability of the interior Neumann problem.

4. The Exterior Neumann Problem

In the two-dimensional case the exterior problem can be reduced to the interior problem. (This cannot be done in the multidimensional case.) We remark that under the condition $\partial u / \partial n = 1$ on the boundary circle $r = 1$, the interior problem cannot be solved, but the exterior problem has the solution $\ln(1/r)$. Any function on the boundary can be represented as the sum of a constant and a function with zero integral over the boundary. By inversion the exterior problem can be reduced to the interior problem: the derivative along the exterior normal becomes the derivative along the interior normal. This is true for an arbitrary region: inversion is a conformal mapping, and the normal to the boundary line becomes the normal to its image under inversion.

The final result is as follows. Without Condition 4 the solution is not unique; under Condition 4 the solution is unique and exists if and only if $\oint_{\partial G} f \, ds = 0$.

The same is true in higher dimensions. Here $s = 1$, $k = 1$, $k - s = 0$.

12.2. Existence and Uniqueness of Solutions

The proofs of all these results on existence and uniqueness of the solutions of boundary-value problems are constructed by the following method. We seek a solution in the form of a potential with an unknown density. An integral equation is obtained for this density and shown to be solvable.

Consider, for example, the interior Dirichlet problem. We seek a solution in the form of a double-layer potential with an unknown density ρ on the boundary. We denote the value of the potential on the boundary by $A\rho$, where A is a linear (integral) operator on the space of functions on the boundary. By the potential jump theorem for the double layer, the limiting value of the potential inside the region is $A\rho + \lambda\rho$, where λ is a constant depending only on the dimension. (It equals π in the two-dimensional case and 2π in the three-dimensional case.) The Dirichlet boundary condition $u\big|_{\partial G} = f$ therefore assumes the form of the equation $(A + \lambda E)\rho = f$ for the density ρ.

If the space of densities were finite-dimensional, we would have the solution immediately: $\rho = (A + \lambda E)^{-1}$. The existence of the inverse operator is guaranteed then by the absence of a nonzero solution in the homogeneous equation.

The uniqueness of the solution of the equation $(A + \lambda E)\rho = 0$ follows from the maximum principle, since a function that is continuous in the closure of the region G and harmonic in its interior and vanishes on the boundary must be identically zero. Therefore if the space of functions ρ were finite-dimensional, the proof of the existence of a solution of the Dirichlet problem would be proved.

Actually this space is infinite-dimensional, but the operator A can be approximated well enough by finite-dimensional operators to preserve the conclusion.

The reason the operator A is nearly finite-dimensional is as follows. The operator maps a density ρ to the value of a potential on the boundary. Consider a point x on the boundary, as shown in Fig. 12.4.

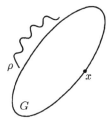

Fig. 12.4. A high harmonic makes only a small contribution to the value of the potential at a boundary point

We expand the density in a Fourier series. A high harmonic of this series oscillates rapidly. Physically this indicates the presence of nearby charges with opposite signs. They act oppositely to each other, so that such a harmonic makes only a small contribution to the result at the point x. Therefore the operator A can be well approximated by finite-dimensional operators. As a result, the index of the operator $A + \lambda E$ equals zero, as for an operator on a finite-dimensional space, although this operator acts on the infinite-dimensional space of densities.

In the multidimensional case for a spherical boundary one can use expansions in spherical functions instead of Fourier expansions. For other boundary manifolds local expansions analogous to Fourier series can be used near each point.

With that we conclude our brief discussion of the questions of existence and uniqueness of solutions of the basic boundary-value problems for Laplace's equation. A detailed proof of the results stated here requires the machinery of the theory of Fredholm integral equations, which we do not have the time to develop in these lectures. A brief exposition of this theory can be found, for example, in the textbook of G. E. Shilov *An Introduction to the Theory of Linear Spaces*.

12.3. Linear Partial Differential Equations and Their Symbols

Let us return to the general theory of differential equations. I begin by recalling the general concept of linearization. Let us start with an ordinary differential equation given by a vector field v in the phase space. Let x_0 be an equilibrium point of the differential equation $\dot{x} = v(x)$, that is, $v(x_0) = 0$. Then in a neighborhood of x_0 one can pose the problem of small oscillations about the equilibrium position, which are described by the linearized system: $\dot{y} = Ay$, where $A = \frac{\partial v}{\partial x}\big|_{x=x_0}$, as shown in Fig. 12.5.

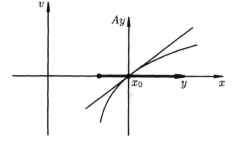

Fig. 12.5. Linearization in a neighborhood of an equilibrium point

Replacing the equation $\dot{x} = v(x)$ by the equation $\dot{y} = Ay$ is called *linearizing* the system at x_0. The linearized vector field "lives" in the tangent space to the original phase space at the equilibrium point. It is independent of the coordinate system used in computing the Jacobian matrix $\frac{\partial v}{\partial x}\big|_{x_0}$.

In the case of partial differential equations the problem of small oscillations of a continuous medium (a string, a membrane, and the like) originates in this way. It is the linearization of the corresponding equation in the dynamics of continuous media, whose phase space is the infinite-dimensional space of positions and velocities of points of the medium. In linearized problems involving small oscillations of a continuous medium the role of the operator A is played by linear partial differential operators, which may be of arbitrarily high order.

The unknown function may be vector-valued rather than scalar-valued. (Picture, for example, a string in three-dimensional space. Its deviation from the equilibrium position at each point is a vector having two components, which are the coordinates in the direction perpendicular to the string.)

To describe such situations in general one must consider the vector bundle whose base is the manifold of independent variables. The values of the unknown functions ("fields," as the physicists call them) define a section of this bundle. In the local theory one can write everything in coordinates and assume that $x \in \mathbb{R}^n$ and $u \in \mathbb{R}^r$. In physics the components of a vector u are called fields; here r is the number of fields and $n = D + 1$, where D is the physical dimension. The extra 1 is reserved for time.

A linear differential operator in the coordinate system $(x_1, \ldots, x_n)$ has the form of a polynomial in the operators $\partial/\partial x_i$ with coefficients depending on x, that is, the form

$$P(\partial, x) = \sum_{|\alpha|=0}^{m} a_\alpha(x)\partial^\alpha ,$$

where

$$\alpha = (\alpha_1, \ldots, \alpha_n) \in \mathbb{Z}_+^n , \quad |\alpha| = \alpha_1 + \cdots + \alpha_n , \quad \partial^\alpha = \frac{\partial^{|\alpha|}}{\partial x_1^{\alpha_1} \cdots \partial x_n^{\alpha_n}} .$$

The number $m = \max |\alpha|$ is called the *order* of the operator. Such an operator acts on scalar-valued functions ($r = 1$). If there are several fields ($r > 1$), then "matrix-valued" differential operators occur with values in the vector bundle over the same base as the original fields. The dimension of a fiber is the same as the number of original fields, namely r. That is, the number of equations equals the number of unknown functions. In local coordinates the operator $u \mapsto v$ can be written as a system of equations:

$$v_1 = P_{11}(\partial, x)u_1 + \cdots + P_{1r}(\partial, x)u_r ,$$

$$\cdots\cdots\cdots\cdots\cdots\cdots\cdots\cdots\cdots\cdots\cdots$$

$$v_r = P_{r1}(\partial, x)u_1 + \cdots + P_{rr}(\partial, x)u_r .$$

If we denote the whole matrix of operators $P_{jl}(\partial, x)$ by $\mathcal{P}(\partial, x)$, we obtain the matrix notation $v = \mathcal{P}(\partial, x)u$. Here $\mathcal{P}$ is a matrix of polynomials in $\partial/\partial x_i$ with coefficients that depend on x.

Example. The Cauchy–Riemann equations have the form

$$0 = \begin{pmatrix} \partial_x & -\partial_y \\ \partial_y & \partial_x \end{pmatrix} \begin{pmatrix} u \\ v \end{pmatrix}.$$

Here $r = 2$, the matrix is skew-symmetric, the degree of the polynomials is $m = 1$, and the coefficients are constant.

Consider, more generally, the important special case when the differential operators $\mathcal{P}$ are translation invariant with respect to x, that is, all the coefficients $a_\alpha(x)$ are constant.

The origin of this case is as follows. If it is known that the fields vary steadily, in first approximation on the local level one can apply the operation of freezing the coefficients, that is, passing to equations with constant coefficients equal to the values of the given coefficients at the point in question. For slowly varying fields freezing gives a rather good approximation.

Thus, let us consider operators with constant coefficients: $A = \mathcal{P}(\partial)$. As it happens, such equations can be solved, and a purely algebraic theory results. Suppose for simplicity that $r = 1$. (We are considering a single equation with one unknown function.)

Consider the translations $T_s : \mathbb{R}^n \to \mathbb{R}^n$, $T_s x = x + s$. The operators $A = \mathcal{P}(\partial)$ and T act on functions and commute with each other: $AT_s = T_s A$. Hence they have common eigenvectors. What eigenfunctions do the translations have? Let us assume that our functions are complex-valued.

Functions on the circle can be represented by Fourier series. As the segment rolled into a circle grows longer, the frequencies occurring in the representation grow closer together. To be specific, for a segment of length L the frequencies are $\kappa = 2\pi l / L$, as shown in Fig. 12.6.

Fig. 12.6. Closely packed frequencies occur in the Fourier expansion

One should expect that as the line segment is stretched to the entire line the series would become an integral. Then the imaginary exponents form a "continuum size basis": e^{ikx}, and k ranges over the dual space $(\mathbb{R}^n)^*$. (In physics k is called a wave vector.) The functions $e_k = e^{ikx}$ are called (harmonic) planar waves; they are eigenfunctions of the translation operator and differentiation:

$$T_s e_k = \lambda_k e_k , \quad \lambda_k = e^{iks} , \quad \partial^\alpha e_k = (ik)^\alpha e_k .$$

We see that $A e_k = P(ik) e_k$, that is, each harmonic planar wave is an eigenfunction of any linear differential operator with constant coefficients, the eigenvalue being a polynomial in the wave vector.

The computations are analogous in the case of r fields: after multiplying the exponential by the amplitude $w \in \mathbb{R}^r$, we obtain $\mathcal{P}(\partial) w e_k = \big(\mathcal{P}(ik)\big) w e_k$. Here $\mathcal{P}(ik)$ is a matrix whose entries are polynomials.

Exercise. Find this matrix for the Cauchy–Riemann operator.

ANSWER. $\begin{pmatrix} ik_1 & -ik_2 \\ ik_2 & ik_1 \end{pmatrix}$.

For any wave vector k and any differential operator $P(\partial)$ with constant coefficients there are, in general, r eigenvectors w_j of the operator $P(ik)$.

Each of them defines a harmonic planar wave with wave vector k of the form $w_j e_k$, which is an eigenvector of the matrix differential operator $A = \mathcal{P}(\partial)$. Physically, planar waves $w_j e_k$ with a given wave vector k have common fronts, but different "polarizations" $w_j \in \mathbb{R}^r$.

When $r = 1$, we have a single differential operator $P(\partial)$, and the polynomial $P(ik)$ in the wave vector k is called the *symbol* of this operator. The highest-degree term of the symbol is called the *principal symbol*.

For example, the symbol of the Laplacian in $\mathbb{R}^n$ is $-k_1^2 - \cdots - k_n^2$ and coincides with its principal symbol. The symbol is usually denoted by the letter σ. For example, for the Laplacian $\sigma = -k^2$.

For a matrix operator the matrix of symbols is called the *matrix symbol* of the operator.

The *symbol of the system* is the determinant of the matrix symbol. For example, the symbol of the Cauchy–Riemann is

$$\det \begin{pmatrix} ik_1 & -ik_2 \\ ik_2 & ik_1 \end{pmatrix} = -k_1^2 - k_2^2 \,,$$

that is, it coincides with the symbol of the Laplacian on the plane.

The *principal symbol of a system* is the principal homogeneous part of its symbol. In wave physics the corresponding concept is usually called the dispersion relation.

Remark. The principal symbol is invariantly determined, that is, it is independent of the coordinate system used to construct it. It is a homogeneous polynomial in the components of the cotangent vector k of the space of independent variables. Invariance of the principal symbol means that its values on each cotangent vector is independent of the coordinate system. The complete symbol is not invariant. For example, terms of first order can be added after a change of variables in the Laplacian. One can see that the principal symbol is invariant from the following considerations. (For simplicity we limit ourselves to the scalar case $r = 1$.) We apply the operator to the planar wave e_k:

$$P(\partial, x)e_k = \sigma_m(k) + \sigma_{m-1}(k) + \cdots,$$

where the coefficients of the polynomials may depend on x. Let $k \to \infty$. If $\sigma_m \neq 0$, this is the principal term of the asymptotics of the value of the operator on a high-frequency planar wave. Making a smooth change of variable x in $e_k(x)$, it is easy to verify that this asymptotics depends only on the vector k, and not on the coordinate system.

The algebraic properties of the principal symbol exert a decisive influence on the properties of the corresponding differential equation, and in this sense the theory of differential equations can be regarded as a branch of algebraic geometry.

Example. Suppose the function f on the manifold can be expanded in a series of exponentials (for example, on the circle or the torus): $f = \sum f_k e_k$. We seek a solution of the equation $\Delta u = f$ in the form of a series $u = \sum u_k e_k$. We obtain $-k^2 u_k = f_k$, $u_k = -f_k/k^2$. It is obvious that the series $\sum u_k e_k$ converges; moreover the smoothness of u has increased by two orders compared with that of f.

The same reasoning goes through for any linear operator of order m with constant coefficients if $|\sigma_m(k)| \geq C|k|^m$.

Such an operator is said to be *elliptic*. We see that the infinite-dimensionality of the problem actually disappears due to ellipticity. The solution method is purely algebraic.

An operator with *variable* coefficients is *elliptic* if it becomes elliptic under freezing of the coefficients at each point.

Example. The Laplacian div grad on any Riemannian manifold (for example, on the surface of a sphere), is elliptic. (Prove this!)

For an elliptic operator $P(\partial, x)$ with variable coefficients the preceding formulas do not give an exact solution of the equation $Pu = f$. However, by freezing the coefficients, one can obtain a very good approximation to the solution of the elliptic equation $Pu = f$ on a compact manifold.

Choosing f to be "wave packets" like $f = e^{-sx^2} e_k(x)$, one can use the approximate solution $u = f/\sigma_m(k)$ when k is large. The rapid decrease of u as $k \to \infty$ is guaranteed because the problem is "almost finite-dimensional." Because of this property the theory of elliptic equations and boundary-value problems is nearly as close to finite-dimensional linear algebra as the corresponding theory for the Laplacian, with which we became acquainted earlier, even in the case of variable coefficients.

If the operator is not elliptic (for example, the wave operator), this method does not go directly through, but it can be modified.

Problem 1. Find the manifold of zeros of the principal symbol of the wave equation

$$\frac{\partial^2 u}{\partial t^2} = \frac{\partial^2 u}{\partial x^2} + \frac{\partial^2 u}{\partial y^2} .$$

ANSWER. This is a quadratic cone in $\mathbb{R}^3$ (the "light cone").

Problem 2. Suppose the quadratic form $\sum a_{pq} k_p k_q$ is positive definite. For which values of the velocity c and the wave vector k does the wave equation

$$\frac{\partial^2 u}{\partial t^2} = \sum a_{pq} \frac{\partial^2 u}{\partial x_p \partial x_q}$$

have a solution in the form of a planar wave $e^{\omega t - kx}$ moving in the direction of the vector k with velocity c?

Problem 3. Find the manifold of zeros of the principal symbol of the system of Maxwell equations
   a) in a vacuum:

$$\frac{\partial E}{\partial t} = \operatorname{curl} H , \qquad \frac{\partial H}{\partial t} = - \operatorname{curl} E ,$$

   b) in a homogeneous nonisotropic medium.

HINT. See the book of Courant and Hilbert, *Equations of Mathematical Physics.*

Problem 4. Find the manifold of zeros of the principal symbol of the system of Dirac equations for four complex-valued functions u_j of the four variables x_i:

$$\sum_{k=1}^{4} \alpha_k (\partial/\partial x_k - a_k) u - \beta b u = 0 ,$$

where

$$\alpha_1 = \begin{pmatrix} 0 & 0 & 0 & 1 \\ 0 & 0 & 1 & 0 \\ 0 & 1 & 0 & 0 \\ 1 & 0 & 0 & 0 \end{pmatrix} , \qquad \alpha_2 = \begin{pmatrix} 0 & 0 & 0 & -i \\ 0 & 0 & i & 0 \\ 0 & -i & 0 & 0 \\ i & 0 & 0 & 0 \end{pmatrix} ,$$

$$\alpha_3 = \begin{pmatrix} 0 & 0 & 1 & 0 \\ 0 & 0 & 0 & -1 \\ 1 & 0 & 0 & 0 \\ 0 & -1 & 0 & 0 \end{pmatrix} , \qquad \alpha_4 = \begin{pmatrix} -1 & 0 & 0 & 0 \\ 0 & -1 & 0 & 0 \\ 0 & 0 & -1 & 0 \\ 0 & 0 & 0 & -1 \end{pmatrix} ,$$

$$\beta = \begin{pmatrix} 1 & 0 & 0 & 0 \\ 0 & 1 & 0 & 0 \\ 0 & 0 & -1 & 0 \\ i & 0 & 0 & -1 \end{pmatrix} .$$

The general outline of the theory of propagation of waves defined by systems of linear partial differential equations is as follows.

The set of zeros of the principal symbol is a field of cones in the cotangent spaces of space-time. Passing to the projectivization (or spherization), we obtain the "Fresnel hypersurface" of retardations in manifolds of contact elements of space-time.

The geometric optics (that is, the theory of rays and fronts) defined by this hypersurface (see Lecture 2) describes approximately (under some conditions) short-wave asymptotics (also called quasiclassical in physics), that is, the behavior of waves whose lengths are small in comparison with the geometric dimensions of the system, for example, compared with distances at which the coefficients of the system vary noticeably.

A fundamental condition is that of *hyperbolicity*, which consists of the following. Consider an algebraic hypersurface of degree d in m-dimensional real projective space. The hypersurface is said to be *hyperbolic* in relation to a point if each real line passing through this point intersects the hypersurface in d *real* points. If these points are pairwise distinct, the surface is *strictly hyperbolic*.

Example. An ellipse is strictly hyperbolic relative to its interior points, nonstrictly hyperbolic relative to its boundary points, and not hyperbolic relative to its exterior points.

The condition of hyperbolicity of a system means that the hypersurface of retardations in the projective space of contact elements at each point of space-time must be hyperbolic relative to the time point.

Here the time point is the projectivization of the vector dt of the cotangent space to space-time. (The corresponding contact element is the tangent hyperplane of an isochrone.)

A strictly hyperbolic surface of even degree $2n$ consists of n ovaloids diffeomorphic to a sphere and situated one inside another from the one closest to the time point to the one most remote from it. Physically, these components correspond to the different "modes" or types of waves that can propagate in the given medium. For example, in an elastic medium there are longitudinal and transverse waves. Longitudinal and transverse waves propagating in the same direction have, in general, different velocities.

The ovaloid closest to the time point corresponds to the slowest wave, the next to a faster wave, and so on, up to the one that arrives first, the fastest wave, which corresponds to the outermost ovaloid. This is clear from the order in which the points where the corresponding cones intersect the vertex O with the timelike world line $q = c$ located alongside O. The point of intersection nearest to the isochrone corresponds to the smallest arrival time t of the perturbation from O at c, and consequently to the fastest wave.

It should be kept in mind that two waves of a given type correspond to each direction; one propagates forward, the other backward. (The corresponding wave vectors are oppositely directed.)

Example. For the wave equation $2n = 2$ there is one mode, and to each direction in physical space there correspond exactly two waves, propagating in opposite directions.

The basis of the short-wave asymptotics and, in particular the transition from the "physical" optics of the wave equation to the geometric optics of the eikonal equation is beyond the scope of these lectures. I remark only that a systematic implementation of this program leads to the appearance of interesting topological invariants in the asymptotic formulas, the so-called Maslov indices, which describe "the loss of one-fourth of the wave when a ray passes through a caustic" and which manifest themselves in quantum mechanics as the correction of $1/2$ in the Bohr–Sommerfeld quantum condition. The general statement of the quantization conditions thus leads to a topological object, the Maslov characteristic class on Lagrangian submanifolds of the symplectic phase space.

Problem. Extend Weyl's formula (the asymptotics of the number of eigenfunctions with eigenvalues less than E) to the case of hyperbolic systems.

Appendix A

The Topological Content of Maxwell's Theorem on the Multifield Representation of Spherical Functions

In this appendix we present the proof of Maxwell's theorem on the multifield representation of spherical functions. In particular it is proved that the functions that admit such a representation form a vector space. This is the proposition that was left unproved above (Lecture 11, p. 113).

Simultaneously we shall prove a number of interesting topological and algebraic corollaries of Maxwell's theorem, which seem to have been proved first by Sylvester in a little-known note that also contains both the basic ideological principle of the treatise of Bourbaki and a warning against the dangers of abusing the formalization of mathematics.

We recall the statement of the theorem (pp. 102 and 112).

Theorem 1. *The restriction to the sphere of an nth-order derivative of the function $1/r$ along n constant (translation-invariant) vector fields in $\mathbb{R}^3$ coincides with a spherical function of degree n. Any nonzero spherical function of degree n can be obtained in this way using some set of n nonzero vector fields. For a given function these fields are uniquely determined (up to a nonzero constant factor and a permutation of the fields).*

The spherical functions of degree n form a vector space of dimension $2n+1$.

The set of functions represented by the multifield construction described in the theorem appears a priori to be essentially nonlinear. It follows from the theorem that the range of the corresponding multilinear mapping is a *linear* manifold. The uniqueness asserted in the theorem can be restated in purely topological terms.

Theorem 2. *The configuration space of n (possibly coincident) indistinguishable points on the real projective plane (that is, the nth symmetric power $\mathrm{Sym}^n(\mathbb{R}P^2)$) is diffeomorphic to the real projective space of dimension $2n$:*

$$\mathrm{Sym}^n(\mathbb{R}P^n) \approx \mathbb{R}P^{2n} \ .$$

The symmetric powers of nonorientable surfaces were computed (independently of Maxwell) by J. Dupont and G. Lusztig [4].

Remark. Theorem 2 is related to Viète's projective theorem

$$\mathrm{Sym}^n(\mathbb{C}P^1) \approx \mathbb{C}P^n$$

and is in a sense the quaternion version of that theorem.

By regarding the Riemann sphere $\mathbb{C}P^1$ as a two-sheeted covering of real projective space, we obtain as a corollary an algebraic mapping $r : \mathbb{C}P^n \to \mathbb{R}P^{2n}$ of multiplicity 2^n, which is a higher-dimensional generalization of the classical theorem asserting that

$$\mathbb{C}P^2/\mathrm{conj} \approx S^4 .$$

A.1. The Basic Spaces and Groups

Consider the n-dimensional arithmetical quaternion space $\mathbb{H}^n = \bigoplus \mathbb{H}_p^1$ with its usual i-complex structure $(\mathrm{i}(a\mathrm{e} + b\mathrm{i} + c\mathrm{j} + d\mathrm{k}) = a\mathrm{i} - b\mathrm{e} + c\mathrm{k} - d\mathrm{j})$. Left multiplication by j operates on $\mathbb{H}_p^1$ while preserving complex lines. It maps each line to its Hermitian-orthogonal line and acts on $\mathbb{C}P_p^1 = (\mathbb{H}_p^1 \setminus 0)/\mathbb{C}^*$ as an antiholomorphic involution σ_p having no fixed points.

Consider the Coxeter group $B(n)$, which operates on the product $(\mathbb{C}P^1)^n$ by permuting factors and applying the mappings σ_p on certain factors.

Theorem 3. *The space of orbits of the action of $B(n)$ on $(\mathbb{C}P^1)^n$ is $2n$-dimensional real projective space:*

$$(\mathbb{C}P^1)^n/B(n) \approx \mathbb{R}P^{2n} .$$

According to Viète's theorem, the space of orbits of the permutation group $S(n)$ is complex projective space:

$$(\mathbb{C}P^1)^n/S(n) = \mathrm{Sym}^n(\mathbb{C}P^1) = \mathbb{C}P^n .$$

Thus we have obtained a natural mapping $r : \mathbb{C}P^n \to \mathbb{R}P^{2n}$, taking the orbit ξ of the subgroup $S(n)$ into the orbit $r(\xi)$ of $B(n)$ containing ξ.

The group $B(n)$ also contains the interesting subgroup $\mathbb{Z}_2 \times S(n)$, consisting of permutations and permutations together with the antiholomorphic involutions σ_p on *each* factor. The product of involutions σ_p operates on $(\mathbb{C}P^1)^n/S(n)$ as an involution $\sigma \in \mathbb{Z}_2$.

The group embeddings $S(n) \to \mathbb{Z}_2 \times S(n) \to B(n)$ generate mappings of the spaces of orbits

$$(\mathbb{C}P^1)^n/S(n) \xrightarrow{\alpha} (\mathbb{C}P^1)^n/(S(n) \times \mathbb{Z}_2) \xrightarrow{\beta} (\mathbb{C}P^1)^n/B(n)$$

of multiplicities 2 and 2^{n-1} respectively.

The involution $\sigma : (\mathbb{C}P^1)^n/S(n) \to (\mathbb{C}P^1)^n/S(n)$ is the pre-image $\alpha^{-1}(\cdot)$.

Theorem 4. *For even n the involution σ operates on $(\mathbb{C}P^1)^n/S(n) \approx \mathbb{C}P^n$ as the complex conjugation* conj.

Thus, for even n we obtain real algebraic mappings

$$\mathbb{C}P^n \xrightarrow{\alpha} \mathbb{C}P^n/\text{conj} \xrightarrow{\beta} \mathbb{R}P^{2n}$$

of multiplicities 2 and 2^{n-1} respectively.

Remark. When $n = 2$, the second space is smooth (see, for example, [1] and [2]; this was probably known even before the publication of [2]):

$$\mathbb{C}P^2/\text{conj} \approx S^4 \,.$$

In this case the multiplicity of β is 2.

The involution representing the two pre-images operates on S^4 as the antipodal involution.

I am grateful to S. Donaldson for this remark, which shows that the (strange) theorem of Maxwell is in a certain sense a multidimensional analog of the (no less strange) theorem that $\mathbb{C}P^2/\text{conj} \approx S^4$.

We shall show below that Maxwell's theorem provides an explicit formula for a diffeomorphism $\mathbb{C}P^2/\text{conj} \to S^4$.

Remark. In the majority of cases when we assert that two manifolds "coincide" we shall exhibit explicitly only a real algebraic homeomorphism between these manifolds. The pedantic verification that these homeomorphisms can be smoothed is in some cases left to the reader (see, however, Sect. A.4).

A.2. Some Theorems of Real Algebraic Geometry

Consider a real homogeneous polynomial f of degree n in the three variables (x, y, z). Maxwell's theorem has the following strange algebraic corollary.

Theorem 5. *Any real homogeneous polynomial f of degree n has a unique representation as the sum of two such polynomials, one of which is the product of n linear real factors, and the other is divisible by $x^2 + y^2 + z^2$.*

In particular, any real algebraic curve of degree n in the metric projective plane determines n real "principal axes" that are invariantly connected with the curve.

PROOF. The real equation $x^2 + y^2 + z^2 = 0$ defines a real curve S in $\mathbb{C}P^2$ having no real points. This curve, which is called an imaginary circle, is rational and is a topological sphere. Complex conjugation maps S into itself and operates on it as an antiholomorphic involution having no fixed points. (It is the usual antipodal involution of S^2.)

The equation $f = 0$ defines a real algebraic curve K of degree n in $\mathbb{C}P^2$. The complex conjugation conj $: \mathbb{C}P^2 \to \mathbb{C}P^2$ maps K into itself and permutes the $2n$ points of the intersection of K with S.

Each pair of conjugate points of the intersection determines the line joining them. (The points are distinct, since conj has no fixed points on S^2.) This line is real (since conj permutes two points belonging to it) and can be defined by an equation $ax + by + cz = 0$ with real coefficients.

Consider the product g of n real linear functions obtained in this way. We shall show that f is proportional to g along S.

The homogeneous polynomial f equals 0 at the $2n$ common points of the curves K and S. The curve S is rational. Let us choose a rational parameter t (for example, $x = 2t$, $y = t^2 - 1$, $z = \mathrm{i}(t^2 + 1)$). Choosing coordinates in a suitable way, we can exclude the case when $t = \infty$ is one of the points of intersection of S with K. The polynomials $f\big(x(t), y(t), z(t)\big)$ and $g\big(x(t), y(t), z(t)\big)$ of degree $2n$ have $2n$ common roots, and $g \neq 0$. Consequently, along S we have the equality $f = cg$, $c = \mathrm{const}$.

Thus, the homogeneous polynomial $f - cg$ equals zero on S, that is, it has the form $(x^2 + y^2 + z^2)h(x, y, z)$, where h is a real homogeneous polynomial of degree $n - 2$. The theorem is now proved. $\square$

Remark. The uniqueness of the decomposition $f = cg + (x^2 + y^2 + z^2)h$ can be proved independently of the proof of existence. Assume that a second decomposition $f = c'g' + (x^2+y^2+z^2)h'$ exists. Then $c'g' - cg = (x^2+h^2+z^2)h''$, where g and g' are both products of n linear factors. Suppose $g = 0$ on the line l. On that line the polynomial $c'g'$ has n real roots and equals the right-hand side of the preceding formula, which has at most $n - 2$ real roots. Thus $c'g' \equiv 0$, and $h'' = 0$ on l. Consequently, we can divide all three polynomials by the equation of l and prove uniqueness by induction on n. (Everything is obvious for $n = 1$, since $cg = c'g'$ by virtue of the fact that $h'' = 0$.)

Consider the projective space $\mathbb{R}P^N$ of real algebraic curves of degree n $(N = n(n + 3)/2)$.

The real algebraic curves consisting of n real lines form a closed real algebraic variety T of dimension $2n$ in this projective space (the image of $(\mathbb{R}P^2)^n$ under a multilinear mapping). The curves containing S form the projective subspace $P = \mathbb{R}P^m$, $M = (n-2)(n+1)/2$. In these terms Theorem 5 assumes the following form:

Theorem 6. *The varieties T and P in $\mathbb{R}P^N$ are linked with linking coefficient 1 in such a way that through any point of $\mathbb{R}P^N$ not belonging to the union of T and P there passes a unique real projective line joining T and P. This line intersects T (just as it intersects P) in a single point.*

It is now easy to prove Theorem 2.

PROOF OF THEOREM 2. Consider the space H of dimension $M+1$ containing M-dimensional projective space P. *Any such space intersects T.* Indeed, let us choose a point O in H not belonging to P or T. The line joining O with T and P is contained in H and intersects T.

The point of intersection is unique. Otherwise at a point of H there would be a point of O through which pass two lines joining T and P.

Thus we have constructed a homeomorphism between $T \approx \mathrm{Sym}^n(\mathbb{R}P^2)$ and a manifold of spaces H of dimension $M + 1$ containing P. This last manifold is $\mathbb{R}P^{N-M-1}$, which (at least on the topological level) proves Theorem 2, since $N - M - 1 = 2n$. $\square$

A.3. From Algebraic Geometry to Spherical Functions

The derivatives of a harmonic function along constant vector fields are obviously harmonic functions. Thus all the repeated derivatives of the function $1/r$ are harmonic except at 0. The following lemmas were proved in Lecture 11 (p. 112).

Lemma 1. *The nth derivative of the function $1/r$ along n constant vector fields has the form P/r^{2n+1}, where P is a homogeneous polynomial of degree n.*

Lemma 2. *The homogeneous polynomial P of Lemma 1 is a harmonic function.*

This lemma follows from the classical theorem on inversion, which we now recall. The harmonic function P/r^{2n+1} is homogeneous and of degree $-(n+1)$. On the unit sphere it coincides with P. It follows that the extension of this function from the unit sphere to the whole space as a homogeneous function of degree n is also harmonic. This extension is precisely P.

To prove the theorem on inversion we consider the spherical Laplacian $\tilde{\Delta}$, which is the Laplacian div grad on the unit sphere. We extend it as a homogeneous function of degree k to $\mathbb{R}^m \setminus \{0\}$ in such a way that any homogeneous function F of degree k is mapped by it into a homogeneous function $\tilde{\Delta}F$ of the same degree.

According to Lecture 11 (p. 105) for any homogeneous function F of degree k we have

$$\tilde{\Delta}F = r^2 \Delta F - \Lambda F , \quad \text{where } \Lambda = k^2 + k(m-2) .$$

The following assertions are consequences of this formula.

(i) A bounded harmonic homogeneous function of degree k on the unit sphere in $\mathbb{R}^m$ is an eigenfunction of the spherical Laplacian corresponding to the eigenvalue $-\Lambda$.

(ii) An eigenfunction of the spherical Laplacian corresponding to the eigenvalue $-\Lambda$, when extended from the sphere to a homogeneous function of degree k on $\mathbb{R}^m \setminus \{0\}$, becomes harmonic.

(iii) For any degree of homogeneity k there exists in $\mathbb{R}^m$ a dual degree $k' = 2 - m - k$ such that a homogeneous harmonic function of degree k remains harmonic it we restrict it to the sphere and then extend it as a homogeneous function of the dual degree. For $m = 3$ the duality condition has the form $k + k' = -1$.

In particular, by Lemma 1, P/r^{2n+1} has degree $k = -(1 + n)$, $m = 3$, that is, the dual degree is $k' = n$, from which Lemma 2 follows.

Lemma 3. *Any spherical function of degree n on S^2 can be represented in the form $f(X, Y, Z)(1/r)$, where f is a homogeneous polynomial of degree n and $X = \partial/\partial x$, $Y = \partial/\partial y$, $Z = \partial/\partial z$, and $r^2 = x^2 + y^2 + z^2$.*

PROOF. The space of harmonic homogeneous polynomials of degree n is a vector space. It contains the subspace of harmonic polynomials whose restrictions can be represented as described in Lemma 3 (by Lemmas 1 and 2). This space is obviously rotation invariant. But the representation of $SO(3)$ in the space of spherical functions of degree n is irreducible. (Any function can be obtained from the Legendre polynomial of degree n by using rotations and forming linear combinations of rotated functions.)

Thus, the subspace of Lemma 3 coincides with the whole space of spherical functions of degree n. □

Lemma 4. *Any spherical function of degree n can be represented in the form $f_T(X, Y, Z)(1/r)$, where $f_T = \prod_{i=1}^{n}(\alpha_i X + \beta_i Y + \gamma_i Z)$ is the product of n real linear factors.*

PROOF. According to Theorem 5 there exists a decomposition $f(x, y, z) = f_T(x, y, z) + g(x, y, z)(x^2 + y^2 + z^2)$. Applying this decomposition to the representation of Lemma 3, we obtain

$$f(X, Y, Z)(1/r) = f_T(X, Y, Z)(1/r) + 0 \,,$$

since $X^2 + Y^2 + Z^2 = \Delta$ and $\Delta(1/r) = 0$. Thus, any spherical function has the multifield representation of Theorem 1. □

Lemma 5. *The multifield representation is unique. That is, the polynomial f_T is uniquely determined by the spherical function.*

PROOF. Assume that f_T and f_T' are completely decomposable polynomials such that

$$f_T(X, Y, Z)(1/r) = f_T'(X, Y, Z)(1/r) \,.$$

According to Lemma 1 (applied n times),

$$f_T(X, Y, Z)(1/r) = (cf_T(x, y, z) + r^2 g)r^{-(2n+1)} \,, \qquad c \neq 0 \,,$$
$$f_T'(X, Y, Z)(1/r) = (c'f_T'(x, y, z) + r^2 g')r^{-(2n+1)} \,, \qquad c' \neq 0 \,.$$

Then $f_T(x, y, z) - f_T'(x, y, z) = r^2 h(x, y, z)$. By Theorem 5, however, this can happen only if $f_T = f_T'$. Theorem 1 is therefore proved. □

A.4. Explicit Formulas

Left quaternion multiplication by j maps the vector $(z, w) = z\mathrm{j} + w\mathrm{e}$ of $\mathbb{C}^2 \approx \mathbb{H}^1$ to $(\bar{w}, -\bar{z})$. This vector is Hermitian-orthogonal to the original.

We thus find an explicit formula $t \mapsto -1/\bar{t}$ for an anti-holomorphic involution of $\mathbb{C}P^1$ having no fixed points.

We can use a pair of Hermitian-orthogonal lines in $\mathbb{C}^2$ to parametrize points in $\mathbb{R}P^2$. In that way we deduce Theorem 3 from Theorem 2. Further, we shall write out explicit formulas for the diffeomorphism

$$\mathrm{Sym}^n(\mathbb{R}P^2) \approx \mathbb{R}P^{2n}$$

of Theorem 2, using sets of n pairs of Hermitian-orthogonal lines in $\mathbb{C}^2$ as coordinates on $\mathrm{Sym}^n(\mathbb{R}P^2)$.

We begin with the case $n = 1$.

To each pair (z, w) and $(\bar{w}, -\bar{z})$ of Hermitian-orthogonal vectors in $\mathbb{C}^2$ we assign a quadratic form on the dual space. This form is a product of the linear functionals represented by these vectors:

$$f = (zx + wy)(\bar{w}x - \bar{z}y) = f_0 x^2 + f_1 xy + f_2 y^2 .$$

Here (x, y) are coordinates on the dual plane of $\mathbb{C}^2$. Thus the coefficients of the quadratic form f have the form:

$$f_0 = z\bar{w} , \quad f_1 = w\bar{w} - z\bar{z} , \quad f_2 = -\bar{z}w .$$

We remark that the coefficient f_1 is real-valued, while $f_2 = -\bar{f}_0$. We shall regard f_0 and f_1 as coordinates on the space $\mathbb{R}^3$ of forms f.

Multiplying the original vector (z, w) by a complex number leads to multiplying the resulting vector by the square of the absolute value of this number. Thus we obtain the mapping

$$F : \mathbb{C}P^1 \to S^2 = (\mathbb{R}^3 \setminus \{0\})/\mathbb{R}^+ ,$$

which maps the complex lines from $\mathbb{C}^2$ to real rays in $\mathbb{R}^3$.

The mapping (f_0, f_1) maps the point (z, w) on the complex line to a completely definite point of the ray. For example, choosing $z = t$, $w = 1$, we obtain $f_0 = t$, $f_1 = 1 - t\bar{t}$.

If the original point on the line is normalized by the condition $|z|^2 + |w|^2 = 1$, its image lies on the ellipsoid $|2f_0|^2 + |f_1|^2 = 1$ (which we can also call a sphere, regarding $2f_0$ and f_1 as coordinates).

Thus the Riemann sphere $\mathbb{C}P^1 = S^3/S^1$ maps to the unit sphere in $\mathbb{R}^3$ with coordinates $2f_0$ and f_1 by the diffeomorphism $(2f_0, f_1)$. Regarding $t = z/w$ as a coordinate on $\mathbb{C}P^1$, we obtain a mapping of the t-plane to the unit sphere in $\mathbb{R}^3$, which is precisely the stereographic projection. The formulas that we write out below are in this sense a multidimensional generalization of the stereographic projection.

By replacing the original line in $\mathbb{C}P^2$ by its Hermitian-orthogonal line, we reverse the sign of the mapping F. Indeed, replacing z by $\bar{w}$ and w by $-\bar{z}$ changes the sign before both f_0 and f_1. Thus, F transforms the involution $j : \mathbb{C}P^1 \to \mathbb{C}P^1$ (which maps every line to its Hermitian complement) into an antipodal involution of the sphere S^2 in $\mathbb{R}^3$.

We now apply the analogous construction to the nth symmetric power of the space $\mathbb{R}P^2$. We begin with the nth symmetric power of $\mathbb{C}P^1$. By definition a point of the complex manifold

$$\operatorname{Sym}^n(\mathbb{C}P^1) \approx \mathbb{C}P^n$$

is a (non-ordered) set consisting of n lines in $\mathbb{C}^2$.

Choosing n nonzero vectors (z_k, w_k) and multiplying the corresponding linear forms on the dual plane, we obtain a binary n-form

$$H(x, y) = \prod_{k=1}^{n} (z_k x + w_k y) = h_0 x^n + \cdots + h_n y^n .$$

The coefficients in this form are (homogeneous) coordinates on $\mathbb{C}P^n = \operatorname{Sym}^n(\mathbb{C}P^1)$ (which define a smooth holomorphic structure on this space).

If $w_k \neq 0$, one can set $w_k = 1$. As affine coordinates we obtain the basic symmetric functions of the variables $\{z_k\}$

$$h_0 = \sigma_n(z) , \quad \ldots , \quad h_{n-1} = \sigma_1(z) , \quad (h_n = 1) .$$

In what follows we shall use these σ_k as *local* coordinates on $\operatorname{Sym}^n(\mathbb{R}P^2)$.

We begin with n pairs of Hermitian-orthogonal lines in $\mathbb{C}^2$. We choose a representative of each pair in such a way that neither of the lines chosen coincides with any of those not chosen (at the given point, and consequently also in some neighborhood of it, where our coordinate system is valid). In order to assure that the nondegeneracy condition above holds in repeating the pair, it suffices to choose the same line each time.

We denote by $(z_k, 1)$ the vectors that define the chosen lines $(k = 1, \ldots, n)$. As local coordinates on the real manifold $\operatorname{Sym}^n(\mathbb{R}P^2)$ we shall use (the real and imaginary parts of) the n complex numbers

$$\sigma_1(z_1, \ldots, z_n) , \quad \ldots , \quad \sigma_n(z_1, \ldots, z_n) .$$

The orthogonal lines are defined by the vectors $(1, -\bar{z}_k)$. We construct the symmetrized $2n$-form

$$f = \prod_{k=1}^{n} (z_k x + y) \prod_{k=1}^{n} (x - \bar{z}_k y) = f_0 x^{2n} + \cdots + f_{2n} y^{2n} .$$

We shall see below that the coefficients f_k are polynomials in σ and $\bar{\sigma}$.

Theorem 7. *The mapping $F : \mathbb{C}^n \to \mathbb{R}^{2n+1}$ taking the point $(\sigma_1, \ldots, \sigma_n)$ to $(f_0, \ldots, f_{2n})$ defines (locally) a diffeomorphism of the manifold $\mathrm{Sym}^n(\mathbb{R}P^2)$ into the space $\mathbb{R}P^{2n}$ of rays in $\mathbb{R}^{2n+1}$. In coordinates the mapping F is given by the following explicit formulas:*

$$f_0 = \sigma_n ,$$
$$f_1 = \sigma_{n-1} - \sigma_n \bar{\sigma}_1 ,$$
$$f_2 = \sigma_{n-2} - \sigma_{n-1} \bar{\sigma}_1 + \sigma_{n-2} \bar{\sigma}_2 ,$$

$$\ldots\ldots\ldots\ldots\ldots\ldots\ldots\ldots\ldots\ldots\ldots\ldots\ldots\ldots$$

$$f_n = 1 - \sigma_1 \bar{\sigma}_1 + \sigma_2 \bar{\sigma}_2 - \cdots + (-1)^n \sigma_n \bar{\sigma}_n .$$

PROOF. It is not difficult to see that

$$\prod_{k=1}^{n}(z_k x + y) = \sigma_n x^n + \sigma_{n-1} x^{n-1} y + \cdots + y^n ,$$
$$\prod_{k=1}^{n}(x - \bar{z}_k y) = x^n - \bar{\sigma}_1 x^{n-1} y + \bar{\sigma}_2 x^{n-2} y^2 + \cdots + (-1)^n \bar{\sigma}_n y^n .$$

Multiplying these two polynomials, we obtain (according to F. Aicardi) the formulas given above for the coefficients of the product.

We also find that $f_{2n-k} = (-1)^{n-k} \bar{f}_k$. In particular, the middle coefficient f_n is real-valued.

It remains to be verified that the Jacobian is nonzero in this region. This can be proved without computation. The Jacobian we are interested in has order $2n + 1$. One of the columns is the vector

$$\Phi = (f_0, \bar{f}_0, f_1, \bar{f}_1, \ldots, f_{n-1}, \bar{f}_{n-1}, f_n) .$$

The other $2n$ columns are its derivatives

$$(\partial\Phi/\partial\sigma_1, \partial\Phi/\partial\bar{\sigma}_1, \ldots, \partial\Phi/\partial\sigma_n, \partial\Phi/\partial\bar{\sigma}_n) .$$

We represent this nonholomorphic Jacobian as the value at the point $\tau = \bar{\sigma}$ of the nonholomorphic Jacobian $T(\sigma, \tau)$ defined by the following construction.

Consider the product

$$\prod_{k=1}^{n}(z_k x + y) \prod_{k=1}^{n}(x - w_k y) = F_0 x^{2n} + \cdots + F_{2n} y^{2n} .$$

The coefficients F_k are polynomials in $\sigma_1(z), \ldots, \sigma_n(z)$ and $\tau_1 = \sigma_1(w), \ldots, \tau_n = \sigma_n(w)$. We denote by $\Psi = (F_0, \ldots, F_{2n})$ a vector-valued function of the variables σ and τ, and we consider the determinant of the matrix

$$(\Psi, \partial\Psi/\partial\sigma_1, \partial\Psi/\partial\tau_1, \ldots, \partial\Psi/\partial\sigma_n, \partial\Psi/\partial\tau_n) .$$

This determinant $T(\sigma,\tau)$ is nonzero at the point $\sigma(z)$, $\tau = \sigma(\bar{z})$. Indeed, it follows from our non-coincidence condition that no z_k coincides with any $-1/\bar{z}_l$. Hence, $(\sigma_1,\dots,\sigma_n)$ and $(\tau_1,\dots,\tau_n)$ form a local holomorphic coordinate system on $\mathbb{C}P^{2n} = \mathrm{Sym}^{2n}(\mathbb{C}P^1)$ in some neighborhood of the point in question σ, $\tau = \bar{\sigma}$. That is, the determinant $T(\sigma,\tau)$ is nonzero at that point. But the determinant we wish to prove nonzero coincides with the value $T(\sigma,\bar{\sigma})$, since $f_{2n-k} = (-1)^{n-k}\bar{f}_k$. Thus, we have proved that the mapping $\mathrm{Sym}^n(\mathbb{R}P^2) \to \mathbb{R}P^{2n}$ is a local diffeomorphism.

From Sect. A.2 we know that this mapping is a homeomorphism. We have now proved Theorem 2 completely. $\square$

A.5. Maxwell's Theorem and $\mathbb{C}P^2/\mathrm{conj} \approx S^4$

The explicit formulas in Sect. A.4 also make it possible to construct a diffeomorphism of the sphere S^{2n} into the orbit space described below.

We begin with the complex manifold $(\mathbb{C}P^1)^n$ of ordered sets of n lines in $\mathbb{C}^2$. On this manifold we consider the following smooth (nonholomorphic) action of the Coxeter group $D(n)$. An element of $D(n)$ operates as a permutation of factors together with the replacement of an *even* number of lines by their Hermitian-orthogonal complements.

Theorem 8. $(\mathbb{C}P^1)^n/D(n) \approx S^{2n}$, and the corresponding diffeomorphism is locally given by the formulas of Theorem 7.

PROOF. Permutations do not change the binary $2n$-form f. Replacing a line by its complement changes the sign of the corresponding factor $(z_k x + y) \times (x - \bar{z}_k y)$. Consequently, an even number of such replacements does not change the sign of f (while an odd number does change it). $\square$

It is interesting that the relation $(\mathbb{C}P^1)^n/D(n) \approx S^{2n}$ thereby proved can be regarded as an informal extension of a theorem of Chevalley: *The orbit space of the action of a real $(2n-1)$-dimensional group on $\mathbb{C}^{2n}$ (which must be regarded as a generalization of the Coxeter group) is the smooth real space $\mathbb{R}^{2n+1}$.*

Example. When $n = 2$, we obtain

$$(\mathbb{C}P^1)^2/D(2) \approx S^4 ,$$

where the group $D(2)$, which consists of 4 elements, operates on pairs of lines from $\mathbb{C}^2$ as permutations and (possibly) replacement of both lines by their Hermitian complements.

But $(\mathbb{C}P^1)^2/S(2) \approx \mathrm{Sym}^2(\mathbb{C}P^1) = \mathbb{C}P^2$. Consequently, $(\mathbb{C}P^1)^2/D(2) = \mathbb{C}P^2/(j)$, where (j) is the replacement of both lines by their complements.

Consider the complex manifold $\mathrm{Sym}^n(\mathbb{C}P^1) = \mathbb{C}P^n$ of unordered sets of n lines in $\mathbb{C}^2$. The operation j of replacing each line by a line Hermitian-orthogonal to it, which acts on $\mathbb{C}P^n$, is an (antiholomorphic) involution.

Theorem 9. *For even n the involution $j : \mathbb{C}P^n \to \mathbb{C}P^n$ coincides with complex conjugation (in some coordinates).*

PROOF. The coefficients of the form

$$H_{z,w}(x,y) = \prod_{k=1}^{n}(z_k x + w_k y) = \sum_{k=0}^{n} h_k x^{n-k} y^k$$

are natural (homogeneous) coordinates on $\mathbb{C}P^n = \mathrm{Sym}^n(\mathbb{C}P^1)$. Let us describe the action of j on the coefficients h_k.

Let us transfer the action of j to the dual space:

$$\bar{w}_k x - \bar{z}_k y = \overline{z_k(-\bar{y}) + w_k(\bar{x})} \ .$$

Thus, the transferred form is given by the formula

$$H_{\bar{w},-\bar{z}}(x,y) = \overline{H_{z,w}(-\bar{y},\bar{x})} \ .$$

In terms of the coefficients we obtain the following expression for the transformed form:

$$\overline{\sum_{k=0}^{n} h_k (-\bar{y})^{n-k}(\bar{x})^k} = \sum_{k=0}^{n}(-1)^{n-k}\bar{h}_k x^k y^{n-k} \ .$$

Thus, the action of j on the coefficients of the form H is given by the formula $(jh)_k = (-1)^k \bar{h}_{n-k}$. Since n is even, we also obtain $(jh)_{n-k} = (-1)^k \bar{h}_k$. The required coordinates are $h_k + h_{n-k}$ and $\mathrm{i}(h_k - h_{n-k})$ for even k, and $\mathrm{i}(h_k + h_{n-k})$ and $h_k - h_{n-k}$ for odd k. (Naturally, we never consider $h_{n/2} - h_{n/2}$.) $\square$

When $n = 2$, our results reduce to an explicit formula for the classical diffeomorphism $\mathbb{C}P^2/\mathrm{conj} \approx S^4$. "Maxwell's theorem"

$$(\mathbb{C}P^1)^n/D(n) \approx S^{2n}$$

extends this diffeomorphism to higher dimensions.

A.6. The History of Maxwell's Theorem

Maxwell's own version of this theorem can be found in his major book *Electricity and Magnetism*, Vol. 1, Chap. IX, Sects. 129–133 (pp. 222–233 in [5]). Sylvester criticized his reasoning, saying in [7]:

> I am rather amazed that such a distinguished author has given no attention to the fact that there always exists one and only one *real* system of poles corresponding to a given harmonic ...

With all due respect to the great talent of Professor Maxwell, I must confess that the derivation of the purely analytic properties of spherical harmonics from "Green's theorem" and the "principle of potential energy," ... as done by him, seems to me an incorrect method, of the same type and just as convincing ... as if the rule for extracting the square root were deduced from Archimedes' law of floating bodies.

Sylvester proposed his own approach, which seems to be equivalent to Theorem 5:

The method of poles for representing spherical harmonics invented or developed by Professor Maxwell actually reduces to choosing a suitable canonical form for a ternary quantic, the sum of the squares of whose variable (here, differential operators) equals zero.

Nevertheless, Sylvester was not concerned with developing this question in detail ("being very pressed for time and in haste to catch a boat sailing for Baltimore in 24 hours"). The details of the proof appeared in [6] and, later, in [3].

Sylvester mentioned the connection of his theory of integrals of products of spherical harmonics with Ivory's theorem on the attraction of an ellipsoid and proposed several generalizations of these ideas.

It appears that neither the algebraic nor the philosophical ideas in this *Note* of Sylvester's were understood or developed by the mathematical community. The pages containing the note were not included in his *Collected Papers* in the library of the Paris École Normale Supérieure.

The note contains the following (anti-Bourbaki) paragraph:

It is by no means unusual for a mathematical investigation ... that a part is in some sense larger than the whole – the reason for this striking intellectual phenomenon is that, in regard to mathematics all quantities and connections must be regarded (as experience teaches) as being in a state of continuous change like the flow of a stream.

This general philosophy led him to the conclusion that "... a general proposition should be easier to prove than any special case of it."

From the fact that Sylvester arrived at this last conclusion we can deduce that the truth of this important principle of him (borrowed by Bourbaki nearly a century later) does not entail the necessity of a lamentable petrification of the "flux" of mathematics.

Literature

1. Arnold, V.I.: A branched covering $\mathbb{C}P^2 \to S^4$, hyperbolicity and projectivity topology. Sib. Math. J., **29**, 717–726 (1988).

2. Arnold, V.I.: Distribution of ovals of real plane algebraic curves, involutions of four-dimensional smooth manifolds, and the arithmetic of integral quadratic forms. Funct. Anal. Appl., **5**, 169–176 (1971).

3. Courant, R., Hilbert, D.: Methods of Mathematical Physics, Vol. 1. Interscience, New York, Chap. 7, § 5 (1953).

4. Dupont, J.L., Lusztig, G.: On manifolds satisfying $w_1^2 = 0$. Topology, **10**, 81–92 (1971).

5. Maxwell, J.C.: A Treatise on Electricity and Magnetism, Unabridged 3rd ed., Vol. 1. Dover, 1954.

6. Ostrowski, A.: Die Maxwellsche Erzeugung der Kugelfunktionen. Jahresber. Deutsch. Math.-Verein., **33**, 245–251 (1925).

7. Sylvester, J.J.: Note on spherical harmonics. Phil. Mag., **2**, 291–307 (1876); see also in: Collected Mathematical Papers of J.J. Sylvester, Vol. 3. Cambridge University Press, 37–51 (1909).

Appendix B

Problems

B.1. Material from the Seminars

Problem 1. Given a vector field v with a singular point of focus type (see the figure), find all solutions of the equation $L_v u = 0$.

Problem 2. The vector field v is given by

$$\dot{x} = y \,,$$
$$\dot{y} = x \,.$$

An initial function for the Cauchy problem is prescribed on the line $x = 1$.

a) What is a necessary condition on the initial function $u\big|_{x=1}$ for the existence of a solution of the Cauchy problem?

b) Is the solution unique?

Problem 3. Given the equation

$$u_t + u u_x = 0 \,, \tag{$*$}$$

a) find its characteristics.

b) Show that Newton's equation

$$\frac{\partial^2 \varphi}{\partial t^2} = 0$$

reduces to $(*)$ under the hypothesis that $\varphi(t)$ is the position of a particle on the line and $u(t, x)$ is the velocity of the particle that is at the point x on the line at time t.

Problem 4. a) Investigate the existence and uniqueness of the solution of the equation

$$\left(\frac{\partial f}{\partial x}\right)^2 + \left(\frac{\partial f}{\partial y}\right)^2 = 1$$

with the initial condition $f\big|_{y=x^2} = 0$ in the regions $y \geq x^2$ and $y \leq x^2$.

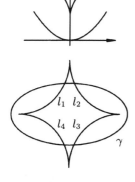

b) Find the curve (see the figure) bounding the region in which the solution is unique. Prove that this curve is the set of centers of curvature of the parabola $y = x^2$.

c) Do the same, but with the initial condition $f\big|_{\frac{x^2}{a^2} + \frac{y^2}{b^2} = 1} = 0$.

d) Consider a small C^3-perturbation of the ellipse. The perturbed curve γ is shown in the figure, and l_1, l_2, l_3, and l_4 are the lengths of the arcs of the enveloping system of normals to γ. Prove that $l_1 + l_3 = l_2 + l_4$.

Problem 5. Does there exist a unique solution of the Cauchy problem

$$x(x^2 + y^2)\frac{\partial u}{\partial x} + y^3\frac{\partial u}{\partial y} = 0, \quad u\big|_{y=0} = 1,$$

in a neighborhood of the point $(x_0, 0)$ on the x-axis?

Homework

Problem 6. Let $\alpha = du - p_1\,dx_1 - p_2\,dx_2$. Prove that no three-dimensional integral surface exists.

Problem 7. Find the characteristics in the $(2n + 1)$-dimensional space of 1-jets of a linear homogeneous equation, a linear nonhomogeneous equation, and a quasi-linear equation.

Problem 8. Investigate the existence and uniqueness of the solution of the problem $yu_x = xu_y$, $u\big|_{x=1} = \cos y$ in a neighborhood of the point $(1, y_0)$.

Problem 9. Find the maximal t for which the solution of the equation

$$\frac{\partial u}{\partial t} + u\frac{\partial u}{\partial x} = \sin x, \quad u\big|_{t=0} = 0$$

can be extended over $[0, t)$.

The Wave Equation, Waves, the Korteweg–de Vries Equation

Problem 10. Find the characteristics of the wave equation

$$\frac{\partial^2 u}{\partial t^2} = c^2 \frac{\partial^2 u}{\partial x^2} , \quad x \in \mathbb{R} .$$

Problem 11. What form does the wave equation assume if the characteristics are used as coordinates?

Problem 12. Find the general form of the solution of the wave equation.

Problem 13. The initial condition $u\big|_{t=0} = u_0(x)$ is prescribed for the wave equation. Find the solution with a given initial condition. Is it unique?

Problem 14. Transform the Korteweg–de Vries (KdV) equation $u_t = 6uu_x - u_{xxx}$ into an ordinary second-order differential equation using the substitution $u = \varphi(x - ct)$ (we are seeking a solution in the form of a moving wave).

Homework

Problem 15. Consider the change of coordinates $x^i = a^{ij}y_j$. Express $\dfrac{\partial u}{\partial x^i}$ in terms of $\dfrac{\partial u}{\partial y^i}$ and $\dfrac{\partial^2 u}{\partial x^i \partial x^j}$ in terms of $\dfrac{\partial^2 u}{\partial y^i \partial y^j}$, and conversely.

Problem 16. Draw the phase portrait of the equation

$$\ddot{\varphi} = 3\varphi^2 + C\varphi + K$$

(the KdV equation after the change of variables $u = \varphi(x - ct)$).

Problem 17. On the phase plane (see Problem 16) find the solution in the form of a wave for which $t \to \pm\infty$, $\varphi \to 0$.

The Wave Equation

Problem 18. Prove d'Alembert's formula

$$u(t, x) = \frac{1}{2}\left(\varphi(x - at) + \varphi(x + at)\right) + \frac{1}{2a} \int\limits_{x-at}^{x+at} \psi(y)\,\mathrm{d}y ,$$

which gives the solution of the equation of the vibrating string $u_{tt} = a^2 u_{xx}$ with initial conditions $u(x, 0) = \varphi(x)$, $u_t(x, 0) = \psi(x)$.

The following problems (Nos. 19–24) involve a semi-infinite string $(x \geq 0)$ with a free left-hand endpoint $(u_x(0, t) \equiv 0)$ or a fixed left-hand endpoint $(u(0, t) \equiv 0)$.

Problem 19. Let the left-hand endpoint of the string be fixed, and suppose initial conditions $\varphi(x)$ and $\psi(x)$ are given for $x \geq 0$. How should the functions φ and ψ be extended to the set $x < 0$ in order that a bounded solution of the resulting Cauchy problem will coincide with the solution of the original Cauchy problem in the region $\{(x,t) : x \geq 0\}$?

Problem 20. Answer the same question for the string with a free left-hand endpoint.

Problems 21–24. Draw movies, that is, representations of the solutions of the Cauchy problem for different values of $t \geq 0$, for the given initial conditions.

21. $\varphi(x)$: 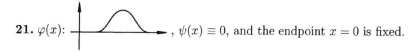 , $\psi(x) \equiv 0$, and the endpoint $x = 0$ is fixed.

22. Same data, but the endpoint $x = 0$ is free.

23. $\varphi(x) \equiv 0$, $\psi(x)$: , and the endpoint $x = 0$ is fixed.

24. Same data, but the endpoint $x = 0$ is free.

Problem 25. Draw movies for the initial conditions as in problems 21–24, but for a string of finite length $(0 \leq x \leq l)$, for the following cases:

a) both endpoints free,

b) both endpoints fixed,

c) one endpoint fixed and the other free.

Homework

Problem 26. Find the general solution of the Cauchy problem for a string of finite length with boundary conditions $u\big|_{x=0} = f(t)$, $u\big|_{x=l} = 0$, if

a) $u\big|_{t=0} = u_t\big|_{t=0}$,

b) $u\big|_{t=0} = \varphi(x)$, $u_t\big|_{t=0} = \psi(x)$.

Problem 27. Suppose that under the conditions of the preceding problem $f(t)$ is periodic with period T.

a) Is the solution $u(x,t)$ a periodic function with respect to t for arbitrary φ and ψ?

b) Verify that there is no periodic solution when $T = (p/q)\tau$, where p and q are integers and $\tau = 2l/a$.

Test

Version 1

1. Solve the Cauchy problem $xu_x + u_y = 0$, $u\big|_{y=0} = \sin x$.

2. For which values of t can the solution of the Cauchy problem $u_t + uu_x = -x^3$, $u\big|_{t=0} = 0$ be extended to the entire half-open interval $[0, t)$?

3. Draw movies for the following problems:

a) $u_{tt} = u_{xx}$, $\varphi \equiv 0$, ψ: , the endpoint $x = 0$ is fixed, the endpoint $x = 1$ is free.

b) $u_{tt} = u_{xx}$, $u\big|_{t=0}$: , $u_t\big|_{t=0} \equiv 0$, $u\big|_{x=0} \equiv 0$.

4. Find the solution of the Cauchy Problem $u_{tt} = u_{xx}$ with initial condition $u\big|_{t=0} = \sin^3 x$, $u_t\big|_{t=0} = \sin x$.

5. Is the series $1 + \dfrac{1}{2}\cos 2x + \dfrac{1}{3}\cos 3x + \cdots + \dfrac{1}{n}\cos nx + \cdots$ the Fourier series of a continuously differentiable function?

6. Let $\{\varphi_k\}$ be an orthonormal system. The series $\sum (f, \varphi_k)\varphi_k$ converges to f in L_2. Express $\|f\|_{L_2}$ in terms of the Fourier coefficients (Parseval's equality).

Version 2

1. Are all solutions of the equation $x\dfrac{\partial u}{\partial x} = y\dfrac{\partial u}{\partial y}$ on $\mathbb{R}^2$ functions of xy?

2. Find the solution of the equation $yuu_x + xuu_y = xy$ whose graph passes through the curve $x = y^2 + u^2 = 1$.

3. Draw movies for the following problems.

a) $u_{tt} = u_{xx}$, $u\big|_{t=0}$: , $u_t\big|_{t=0}$: .

b) $u_{tt} = u_{xx}$, φ: , $\psi \equiv 0$, the endpoint $x = 0$ is fixed, and the endpoint $x = 1$ is free.

4. Solve the Cauchy problem $u_{tt} = u_{xx}$ with the initial conditions $u\big|_{t=0} = \cos x$, $u_t\big|_{t=0} = \cos^3 x$.

5. Is the series $1 + \dfrac{1}{2\ln 2}\cos 2x + \cdots + \dfrac{1}{n\ln n}\cos nx + \cdots$ the Fourier series of a continuously differentiable function?

6. Let $\varphi \in C^{\omega}(S^1)$; more precisely, the function φ is holomorphic in $|\operatorname{Im} z| < \beta$ and $\varphi(z + 2\pi) = \varphi(z)$. Prove that $|\alpha_k| < c e^{-\beta k}$.

Conservative Systems. Lissajous Figures

Problem 28. Consider the equation $\ddot{x} = -\omega^2 x$.

a) Find all solutions.

b) Suppose the solution has the form $A \sin \omega t + B \cos \omega t = C \sin(\omega t + \varphi)$. Prove that $A^2 + B^2 = C^2$.

In what follows we are considering the equation

$$\ddot{x} = -\nabla U(x) .\tag{B.1}$$

Problem 29. Given the potential $U(x_1, x_2) = \dfrac{1}{2}(x_1^2 + x_2^2)$, find all solutions of the equation (B.1) in the form shown in part b) of Problem 28.

Problem 30. Given the potential $U(x_1, x_2) = x_1^2 + 4x_1 x_2 + 4x_2^2$, find all the solutions of the equation (B.1). How do the solutions behave in the (x_1, x_2)-plane?

Problem 31. Prove that the total energy of the conservative system (B.1) is a first integral of this system.

In what follows we consider the potential $U(x) = \dfrac{x_1^2}{2} + \dfrac{\omega^2 x_2^2}{2}$. With a suitable choice of origin for the time t the general solution of (B.1) can be written as

$$x_1 = A_1 \sin t ,$$
$$x_2 = A_2 \sin(\omega t + \varphi) .$$

Problem 32. Let $\omega = 1$, $\varphi = \frac{\pi}{2}$, π, and 0. How will the Lissajous figures look in the rectangle $|x_1| \leq A_1$, $|x_2| \leq A_2$? What will happen if φ varies from 0 to π? Find the coefficients of the similarity transformation of the ellipses $U(x_1, x_2) = E$ and the Lissajous figure when $\varphi = \frac{\pi}{2}$.

Homework

Problem 33. Given the potential $U(x, y) = 5x^2 + 5y^2 - 7xy$. Solve the equation (B.1).

Problem 34. Given the potential $U(x_1, x_2) = \dfrac{x_1^2}{2} + \dfrac{1}{2}\omega^2 x_2^2$. Consider the ellipse $U(x_1, x_2) \leq E$. Let $E_1 = \dfrac{\dot{x}_1^2}{2} + \dfrac{x_1^2}{2}$, $E_2 = \dfrac{\dot{x}_2^2}{2} + \dfrac{\omega^2 x_2^2}{2}$, and $E = E_1 + E_2$. Then x_1 and x_2 lie in the strips $|x_1| \leq \sqrt{2E_1}$, $|x_2| \leq \sqrt{2E_2}$. Prove that the rectangle $|x_1| \leq \sqrt{2E_1} = A_1$, $|x_2| \leq \sqrt{2E_2} = A_2$ is inscribed in the ellipse $U(x_1, x_2) \leq E$.

Problem 35. Prove that if

 a) $\omega = \frac{m}{n}$, the Lissajous figure is a closed curve,

 b) $\omega \notin \mathbb{Q}$, it is not closed.

Problem 36. Prove that if $\omega = n$, there exists a phase φ such that the Lissajous figure is the graph of a polynomial of degree n (the Chebyshev polynomial $p(x) = \cos(n \arccos x)$).

Problem 37. Prove that the ellipse $\dfrac{x^2}{16} + \dfrac{y^2}{25} = 1$ and the arc of the sinusoidal curve $y = 3\cos(x/4)$, $0 \le x \le 8\pi$, have the same length.

Harmonic Functions

Problem 38. Prove that the angle subtended by a line segment from a point in its plane is a harmonic function on the covering of the plane with the endpoints of the interval removed.

Problem 39. Construct a function f harmonic in the disk and assuming given values C_1 and C_2 on the arcs S_1 and S_2, where $S_1 \cup S_2 = S^1$.

Problem 40. Solve the analogous problem for a partition of the circle into n arcs $S_1, \ldots, S_n$ with values $C_1, \ldots, C_n$.

Problem 41. Write Newton's equation of motion for a free particle in polar coordinates.

Homework

Problem 42. Prove that the Lagrangians

$$L_1 = \sqrt{\sum_{i,j} g_{ij} \dot{x}^i \dot{x}^j}, \quad L_2 = \sum_{i,j} g_{ij} \dot{x}^i \dot{x}^j$$

minimize the action on a fixed curve (the shortest curve joining two points).

Problem 43. Compute Δ in polar and spherical coordinates.

Problem 44. Prove that the magnitude of the solid angle subtended from a variable point of space by a fixed closed contour in $\mathbb{R}^3$ is a harmonic function of the variable point on the manifold covering the complement of the curve.

B.2. Written Examination Problems

1995[1]

1 (1). Is the surface

$$z = r^2 + \frac{2}{9} \qquad (*)$$

in Euclidean space smooth? (Here $r^2 = x^2 + y^2 + z^2$.) Is it convex? Find its curvature.

2 (2). Find the value at the origin of a double-layer potential of density 1 distributed on the surface whose equation is $(*)$.

3 (1, 2, 3, 3, 6). Compute the average values over the surface $(*)$ for the following functions: a) z; b) $1/r$; c) z/r^3; d) r^2; e) $1/r^3$.

4 (5). Solve the interior Dirichlet problem with boundary condition $u = 1/r^3$ on the surface $(*)$ for Laplace's equation $\Delta u = 0$.

5 (5). Consider the single-layer potential of density z distributed on the surface $(*)$. Find the average value of this potential over the surface of the sphere $r^2 = 1$.

6 (2, 4). Find the upper and lower bounds of the Dirichlet integral

$$\iiint (\operatorname{grad} u)^2 \, dx \, dy \, dz$$

over the region bounded by the surface $(*)$ on the set of smooth functions in the closure of this region taking the value r^2 on its boundary.

7 (6). Find the value of the solution f of the equation

$$\frac{\partial f}{\partial t} + \operatorname{div}(f \operatorname{grad} u) = f^2$$

for small $|t|$ at the point $(x = y = 0, \; z = 1/2)$ with initial condition $(f \equiv 1$ when $t = 0)$, where u is the solution of Problem 4 in this examination.

HINT. It is not necessary to know the answer to Problem 4 in order to do this. Problem 7 can be solved independently.

---

[1] The number of points awarded for a correct solution of a problem or part of a problem is indicated in parentheses next to the problem number. The recommended criteria for grading in a three-hour examination are: average – 12; good – 16; excellent – 26 (the maximum possible is 40).

1996

1. Given the vector field $v(x,y) = y\frac{\partial}{\partial x} + x\frac{\partial}{\partial y}$ and the function $u\big|_{x=1} = f(y)$, under what condition on f in a neighborhood of the point $(1,0)$ does there exist a solution of the Cauchy problem for the equation $L_v u = 0$? Is it unique?

2. Consider the equation

$$\ddot{x} = -\nabla U(x) , \qquad\qquad (**)$$

where $x = (x_1, x_2)$ and $U(x) = x_1^2 + x_2^2 + ax_1x_2$. For which (real) values of a are all the solutions of the equation $(**)$ periodic?

3. Electric charge is distributed on two lines in $\mathbb{R}^3$: on the line $z = 1$, $y = x$ with density 1 and the line $z = -1$, $y = -x$ with density -1. Find the equipotential surfaces of the field created by these charges.

4. The function v is defined on the sphere $S^2 : x^2 + y^2 + z^2 = 1$ and is harmonic except at the point $N = (0,0,1)$. Let $\mathbb{R}^2$ be the plane $z = 0$ and $p : S^2 \setminus N \to \mathbb{R}^2$ the stereographic projection. Let $u(x,y) = v\big(p^{-1}(x,y)\big)$. Prove that $\int\limits_0^{2\pi} u_r(1,\varphi)\,d\varphi = 0$.

Universitext

Aksoy, A.; Khamsi, M. A.: Methods in Fixed Point Theory

Alevras, D.; Padberg M. W.: Linear Optimization and Extensions

Andersson, M.: Topics in Complex Analysis

Aoki, M.: State Space Modeling of Time Series

Arnold, V.I.: Lectures on Partial Differential Equations

Audin, M.: Geometry

Aupetit, B.: A Primer on Spectral Theory

Bachem, A.; Kern, W.: Linear Programming Duality

Bachmann, G.; Narici, L.; Beckenstein, E.: Fourier and Wavelet Analysis

Badescu, L.: Algebraic Surfaces

Balakrishnan, R.; Ranganathan, K.: A Textbook of Graph Theory

Balser, W.: Formal Power Series and Linear Systems of Meromorphic Ordinary Differential Equations

Bapat, R.B.: Linear Algebra and Linear Models

Benedetti, R.; Petronio, C.: Lectures on Hyperbolic Geometry

Berberian, S. K.: Fundamentals of Real Analysis

Berger, M.: Geometry I, and II

Bliedtner, J.; Hansen, W.: Potential Theory

Blowey, J. F.; Coleman, J. P.; Craig, A. W. (Eds.): Theory and Numerics of Differential Equations

Börger, E.; Grädel, E.; Gurevich, Y.: The Classical Decision Problem

Böttcher, A; Silbermann, B.: Introduction to Large Truncated Toeplitz Matrices

Boltyanski, V.; Martini, H.; Soltan, P. S.: Excursions into Combinatorial Geometry

Boltyanskii, V. G.; Efremovich, V. A.: Intuitive Combinatorial Topology

Booss, B.; Bleecker, D. D.: Topology and Analysis

Borkar, V. S.: Probability Theory

Carleson, L.; Gamelin, T. W.: Complex Dynamics

Cecil, T. E.: Lie Sphere Geometry: With Applications of Submanifolds

Chae, S. B.: Lebesgue Integration

Chandrasekharan, K.: Classical Fourier Transform

Charlap, L. S.: Bieberbach Groups and Flat Manifolds

Chern, S.: Complex Manifolds without Potential Theory

Chorin, A. J.; Marsden, J. E.: Mathematical Introduction to Fluid Mechanics

Cohn, H.: A Classical Invitation to Algebraic Numbers and Class Fields

Curtis, M. L.: Abstract Linear Algebra

Curtis, M. L.: Matrix Groups

Cyganowski, S.; Kloeden, P.; Ombach, J.: From Elementary Probability to Stochastic Differential Equations with MAPLE

Dalen, D. van: Logic and Structure

Das, A.: The Special Theory of Relativity: A Mathematical Exposition

Debarre, O.: Higher-Dimensional Algebraic Geometry

Deitmar, A.: A First Course in Harmonic Analysis

Demazure, M.: Bifurcations and Catastrophes

Devlin, K. J.: Fundamentals of Contemporary Set Theory

DiBenedetto, E.: Degenerate Parabolic Equations

Diener, F.; Diener, M.(Eds.): Nonstandard Analysis in Practice

Dimca, A.: Singularities and Topology of Hypersurfaces

DoCarmo, M. P.: Differential Forms and Applications

Duistermaat, J. J.; Kolk, J. A. C.: Lie Groups

Ma, Zhi-Ming; Roeckner, M.: Introduction to the Theory of (non-symmetric) Dirichlet Forms

Mac Lane, S.; Moerdijk, I.: Sheaves in Geometry and Logic

Marcus, D. A.: Number Fields

Martinez, A.: An Introduction to Semiclassical and Microlocal Analysis

Matoušek, J.: Using the Borsuk-Ulam Theorem

Matsuki, K.: Introduction to the Mori Program

Mc Carthy, P. J.: Introduction to Arithmetical Functions

Meyer, R. M.: Essential Mathematics for Applied Field

Meyer-Nieberg, P.: Banach Lattices

Mikosch, T.: Non-Life Insurance Mathematics

Mines, R.; Richman, F.; Ruitenburg, W.: A Course in Constructive Algebra

Moise, E. E.: Introductory Problem Courses in Analysis and Topology

Montesinos-Amilibia, J. M.: Classical Tessellations and Three Manifolds

Morris, P.: Introduction to Game Theory

Nikulin, V. V.; Shafarevich, I. R.: Geometries and Groups

Oden, J. J.; Reddy, J. N.: Variational Methods in Theoretical Mechanics

Øksendal, B.: Stochastic Differential Equations

Poizat, B.: A Course in Model Theory

Polster, B.: A Geometrical Picture Book

Porter, J. R.; Woods, R. G.: Extensions and Absolutes of Hausdorff Spaces

Radjavi, H.; Rosenthal, P.: Simultaneous Triangularization

Ramsay, A.; Richtmeyer, R. D.: Introduction to Hyperbolic Geometry

Rees, E. G.: Notes on Geometry

Reisel, R. B.: Elementary Theory of Metric Spaces

Rey, W. J. J.: Introduction to Robust and Quasi-Robust Statistical Methods

Ribenboim, P.: Classical Theory of Algebraic Numbers

Rickart, C. E.: Natural Function Algebras

Rotman, J. J.: Galois Theory

Rubel, L. A.: Entire and Meromorphic Functions

Rybakowski, K. P.: The Homotopy Index and Partial Differential Equations

Sagan, H.: Space-Filling Curves

Samelson, H.: Notes on Lie Algebras

Schiff, J. L.: Normal Families

Sengupta, J. K.: Optimal Decisions under Uncertainty

Séroul, R.: Programming for Mathematicians

Seydel, R.: Tools for Computational Finance

Shafarevich, I. R.: Discourses on Algebra

Shapiro, J. H.: Composition Operators and Classical Function Theory

Simonnet, M.: Measures and Probabilities

Smith, K. E.; Kahanpää, L.; Kekäläinen, P.; Traves, W.: An Invitation to Algebraic Geometry

Smith, K. T.: Power Series from a Computational Point of View

Smoryński, C.: Logical Number Theory I. An Introduction

Stichtenoth, H.: Algebraic Function Fields and Codes

Stillwell, J.: Geometry of Surfaces

Stroock, D. W.: An Introduction to the Theory of Large Deviations

Sunder, V. S.: An Invitation to von Neumann Algebras

Tamme, G.: Introduction to Étale Cohomology

Tondeur, P.: Foliations on Riemannian Manifolds

Verhulst, F.: Nonlinear Differential Equations and Dynamical Systems

Wong, M. W.: Weyl Transforms

Druck- und Bindearbeiten: Legoprint, Italien

Manufactured by Amazon.ca
Bolton, ON

37252995R00098